Seaweeds of the Pacific Coast

Common Marine Algae from Alaska to Baja California

by

Jennifer Mondragon and Jeff Mondragon

A SEA CHALLENGERS PUBLICATION
AN IMPRINT OF SHORELINE PRESS
2010

A SEA CHALLENGERS PUBLICATION
AN IMPRINT OF SHORELINE PRESS

Distributed By
Pacific books
P.O. Box 3562
Santa Barbara, CA 93130
(805) 687-8340
pacificbooks@gmail.com

Editor, Dr. Robert Rasmussen

Front Cover
Nereocystis luetkeana (top left), *Macrocystis pyrifera* (top right)
Bossiella (center), *Codium setchellii* (bottom left)
Laminaria and *Phyllospadix* (bottom right)
Cover photographs by Jeff Mondragon

Library of Congress Cataloging-in-Publication Data
Mondragon, Jennifer
Seaweeds of the Pacific Coast : common marine algae from Alaska to Baja California / by Jennifer Mondragon and Jeff Mondragon.
96 p. Includes bibliographic references (2 p.)
ISBN 10: 1-885375-21-2 and ISBN 13: 978-1-885375-21-6
1. Marine algae--Pacific Coast (North America)--Identification. 2. Marine algae--Pacific Coast (North America)--Pictorial works. I. Mondragon, Jeff II. Title.

QK570.5.M66 2010
579.8'8'0979--dc21 2002191158

Shoreline Press
2573 Treasure Drive
Santa Barbara, CA 93105

Printed in Hong Kong through Global Interprint, Santa Rosa, CA, USA
Typography and prepress production by Jeff Mondragon and Jennifer Mondragon, Douglas, AK, and Diana Behrens, Danville, CA, USA.

Table of Contents

Acknowledgments

There were many people who assisted us while we worked on this book and we could not have completed this project without their help! We especially wish to thank the following people:

Berkley White and Jessica Wheeler for repeatedly letting us invade their house. They graciously tolerated our coming and going and provided much-needed, late night conversation and refreshments. Special thanks to Berkley for allowing us to turn the back room of Backscatter Underwater Video and Photo in Monterey, California into an ad hoc wet lab.

Subaquatic Camera Repair in Salinas, California,for keeping our cameras in operating condition during this project, despite our constant efforts to the contrary.

Marc and Jennifer Williams for providing us with floating accommodations in Sitka, Alaska.

Ron and Patty Dick for generously allowing us the use of their boat while we were in Sitka.

Dr. Michael Stekoll for providing access to the University of Alaska Southeast herbarium and allowing us to photograph specimens of *Delesseria* and *Ceramium*.

The late Dr. Thomas DeCew for his encouragement and infectious enthusiasm for phycology.

Lisa Scharf for creating the beautiful artwork in the introduction.

Dr. Robert Rasmussen for editing the book. His many helpful comments and suggestions were invaluable and greatly improved the manuscript.

Photo and Art Credits

All photographs were taken by Jeff Mondragon except for the following:

Dan Richards: *Hesperophycus californicus*
Dan Gotshall: *Rhodymenia pacifica*

Drawings by Lisa Scharf.

Introduction

Scope of this book

We have created this book for those who enjoy working or playing on the shores or in the shallow waters of the west coast of North America, people who will inevitably come in contact with seaweeds and will want to know more about them. From the cold water of the Gulf of Alaska to the warm temperate water of Baja California, there is an incredible diversity of seaweeds. Whether you are a student, teacher, scientist, naturalist or simply have a natural curiosity for the organisms that inhabit the coastal waters, this book will help you recognize and identify the common and widespread seaweeds found along this coast.

What are seaweeds?

Seaweeds are representatives of three highly disparate groups of photosynthetic organisms that inhabit the shores and shallows of the oceans. When the seaweeds are combined with their microscopic peers they are collectively known as the marine algae. When the marine algae are combined with their fresh-water counterparts, they are known simply as algae. In the 18th Century the algae, as a whole, were treated as a formal taxon, Class Algae, within the Kingdom Plantae. Today we know that these photosynthetic marvels are not related closely to the plants or to one another and the term algae (singular, alga) is now a useful common noun with no formal systematic implications.

Major seaweed groups

Despite a lack of genetic relationship, the similarities in algal forms and their common lifestyles make it convenient to study seaweeds as a unit. The study of algae is called phycology and the word is derived from the Greek root *phykos* which means seaweed. Seaweeds fall into 3 main groups: green algae, brown algae, and red algae.

Green Algae

The green algae, or Phylum Chlorophyta, are a very diverse group that contains about 8,000 species. This phylum has evolutionary importance as descendants of the organisms that gave rise to plants. Like plants, green algae have two forms of chlorophyll, *a* and *b*, which they use to capture sunlight to manufacture food. Many green algae are also similar to plants in that their cell walls are composed of cellulose and they use starch as an energy storage molecule. Unlike plants, green algae are primarily aquatic. Most green algae live in fresh water, but there are many marine and some terrestrial species.

Green algae are usually distinguished by their bright grass-green color that is a result of their chloroplasts not being masked by accessory pigments. Members of this phylum can be unicellular, multicellular, colonial, or coenocytic (where the organism is composed of one large cell that has many nuclei and no cross-walls). Green algae are often motile or have motile spores and gametes. Motile cells usually have two or four flagella and the flagella are similar to each other in size and structure.

Brown Algae

The brown algae are represented by about 2,000 species that are almost entirely restricted to marine habitats. All species in this group are multicellular and their morphology varies from tiny, branched filaments to enormous, tree-like kelps with complex anatomy. Brown algae get their name from the brown color of their chloroplasts that, in addition to chlorophyll *a* and *c*, contain brown accessory pigments such as fucoxanthin. The amounts and combinations of their pigments vary and brown algae can be light to dark brown, yellowish, or olive green.

In brown algae, the main reserve products from photosynthesis are the polysaccharide, laminarin, and the alcohol, mannitol. The cell walls consist of cellulose and two groups of compounds, alginates and fucans, that provide flexibility to the brown algae and help prevent desiccation. Alginates have many industrial uses as gels and emulsifiers and are used in the food industry in products such as ice cream and yogurt.

Red Algae

The red algae, classified as the phylum Rhodophyta, include about 6,000 species, which are predominantly marine, with a few freshwater exceptions. The only chlorophyll in red algae is chlorophyll *a* that is usually masked by a red accessory pigment, phycoerythrin. A blue accessory pigment, phycocyanin, is also very common in Rhodophyta. The colors that result from the blend of these pigments are incredibly varied (red algae are not always red!) ranging from whitish-pink, rose-red, yellowish-green, dark brown to purplish black.

Unlike most other algae, the reproductive cells of red algae never have flagella. The food reserve in red algae is the polysaccharide floridean starch. Cell walls of Rhodophyta contain some cellulose and large quantities of mucilaginous material composed of different forms of the sugars galactose and mannose. Two of these compounds, agar and carrageenan, are used in the food, textile and pharmaceutical industries.

Nomenclature

Names have power. When we know the name of an organism we can think about it more clearly and we can communicate our thoughts with others. Seaweed species, unlike fish and bird species, seldom have common names that are widely known, so generally their latinized scientific names are used. The scientific name is composed of two words, for example, *Laminaria farlowii*. The first word in the scientific name is the genus and is capitalized; the second word is the trivial name, or specific epithet, and is entirely lower case. Both words in the species name are always either italicized or underlined.

Naming and classifying organisms is a ongoing process. Scientific names are continually changed and revised as scientists learn more about relationships between species. When name changes occur, the old name becomes a synonym to the new name. For this reason, you will often find the same seaweed described with different names, in books published at different times. In the identification section of this book, to help track some of these taxonomic revisions, we have listed the older scientific names for species that have undergone recent name changes. In addition, each species name is followed by the name of the author who first described the species. If the work of another person has resulted in a change of the original name, the author who first described that species is listed in parentheses and the author of the currently accepted name follows.

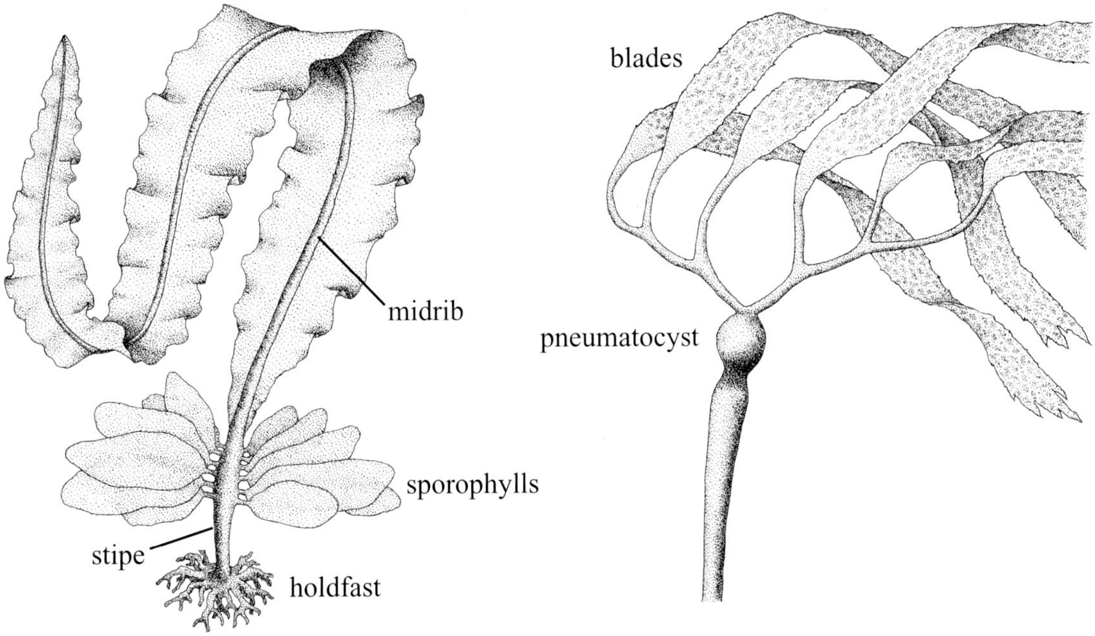

Figure 1. The kelps, *Alaria* and *Pelagophycus*, illustrating various thallus structures.

Seaweed Structures

The body of a seaweed is called the thallus, and there are many terms to describe the various parts of the thallus (Figure 1). At the bottom of the thallus there may be a holdfast that anchors an alga to the substrate. In kelp, there are two main types of holdfasts (Figure 2): holdfasts composed of haptera that are root-like in appearance and discoid holdfasts that are circular or disk-like in shape. Green, red, and small brown seaweeds attach to the substrate in a variety of ways including small discoid holdfasts, rhizoids, and rhizomes.

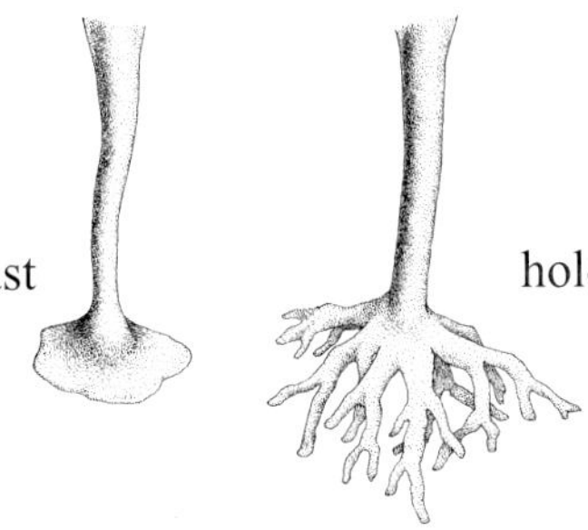

Figure 2. The two primary types of holdfasts found in kelp, discoid and those composed of haptera.

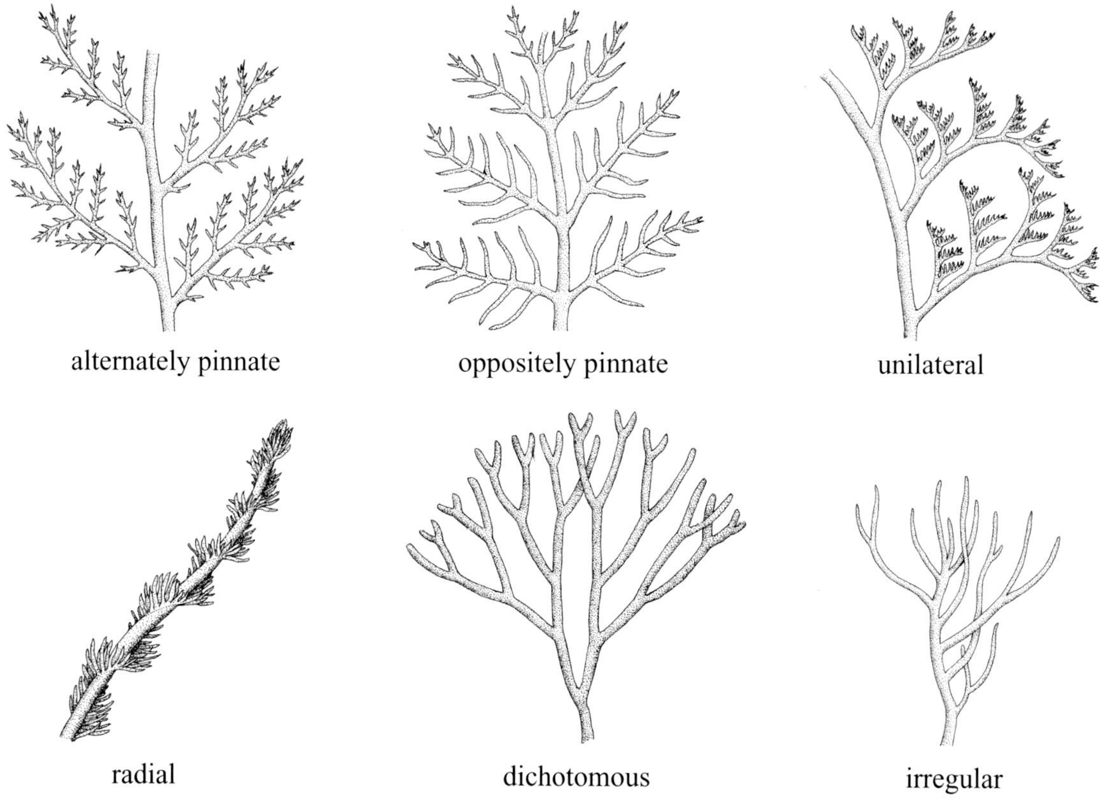

Figure 3. Some of the common branching patterns observed in seaweeds.

Above the holdfast there may be a stem-like portion of the thallus called the stipe. The stipe supports branches and leaf-like structures called blades. A midrib, or thickened longitudinal line, often runs down the length of the blade. Specialized blades that bear sporangia, called sporophylls, may also be attached to the stipe. Some seaweeds also have gas-filled pneumatocysts that serve as floats.

To accurately identify many seaweeds, it is important to be able to differentiate branching patterns (Figure 3). Four of these branching patterns have a main axis that bears lateral branches. Pinnate branching (also called distichous) is feather-like and the main axis bears lateral branches off two sides. Pinnate branching can be either opposite or alternate. In opposite pinnate branching, the lateral branches grow directly across from one another on the main axis. In alternate pinnate branching, the lateral branches are staggered relative to each other on the main axis. Unilateral and radial branching patterns also have lateral branches that arise from a main axis. In unilateral branching, the branches only grow from one side of the main axis. This pattern may be repeated through the subsequent orders of branching, giving each branch a comb-like appearance. In radial branching, as the name implies, branches occur on all sides of the main axis.

Two other types of branching patterns, in which a main axis may or may not be evident, are dichotomous and irregular. Dichotomous branching is characterized by Y-shaped branches of equal length. In contrast, irregular branching has no uniform pattern.

Seaweed Reproduction

Most seaweeds have sexual reproduction in which there are thalli whose cells have only one set of chromosomes (haploid/n). Haploid thalli are called gametophytes because they produce single celled gametes that fuse in an act of fertilization to produce a zygote with two complete sets of chromosomes (diploid/2n). The zygote grows into a thallus, called a sporophyte, whose cells are diploid. Certain cells of the sporophyte undergo meiotic division to produce haploid meiospores . These meiospores grow into gametophytes to repeat the cycle. The gametophytes and sporophytes are known as generations; the recurring sequence of thalli in the life-cycle is called alternation of generations. Classic alternation of generations is found throughout the green and brown seaweeds.

In many instances (*Ulva,* for example, shown in Figure 4) the haploid and diploid thalli are identical in form, or isomorphic. In other seaweeds (*Nereocystis*, for example, Figure 5) one generation, usually the sporophyte, is much larger and more complex than the other, which may be strongly reduced. This is a heteromorphic alternation of generations and is common in kelps and siphonous green algae. In many heteromorphic life cycles, the phases have such dramatic differences in appearance that they were originally described as separate species. It was not until the algae were cultivated in a laboratory that the alternate phases were discovered to be one species! Some phycologists speculate that structural differences in life cycle stages may provide a selective advantage from herbivory or from seasonal variation in habitat.

In a few seaweed genera (*Fucus*, for example, Figure 6) the haploid generation is absent. Meiotic division occurs in the sporophyte and produces gametes. The gametes fuse to form a zygote that germinates into a new sporophyte. This life history pattern is similar to that of humans and the other vertebrate animals.

In some seaweeds the gametes are identical in appearance, isogamy, but differ physiologically. These gametes are small and have various types and numbers of flagella, which, like the flagellum on a human sperm cell, allow them to swim. In some algae the motile gametes are of different sizes, but otherwise similar in appearance, anisogamy. In the larger seaweeds one gamete is large and immobile, an egg, and the other is small and mobile, the sperm. This is termed oogamy. The spores of most green and brown seaweeds are motile, zoospores, and resemble isogametes or sperm.

The red seaweeds (*Polysiphonia*, for example, Figure 7) have a distinctive life history with distinctive reproductive structures. There are no motile cells in the red seaweeds and they are all oogamous. The non-motile males cells, or spermatium, are carried passively to the female reproductive cell, the carpogonium, where fertilization is accomplished.

One or several diploid nuclei result from red algal fertilization. The diploid condition is frequently transferred to other cells known as generative auxiliary cells. Diploid multicellular structures grow from these diploid cells. Since they appear to be the result of germination of a zygote, these diploid structures are often referred to as a new generation, the carposporophyte. The carposporophyte is

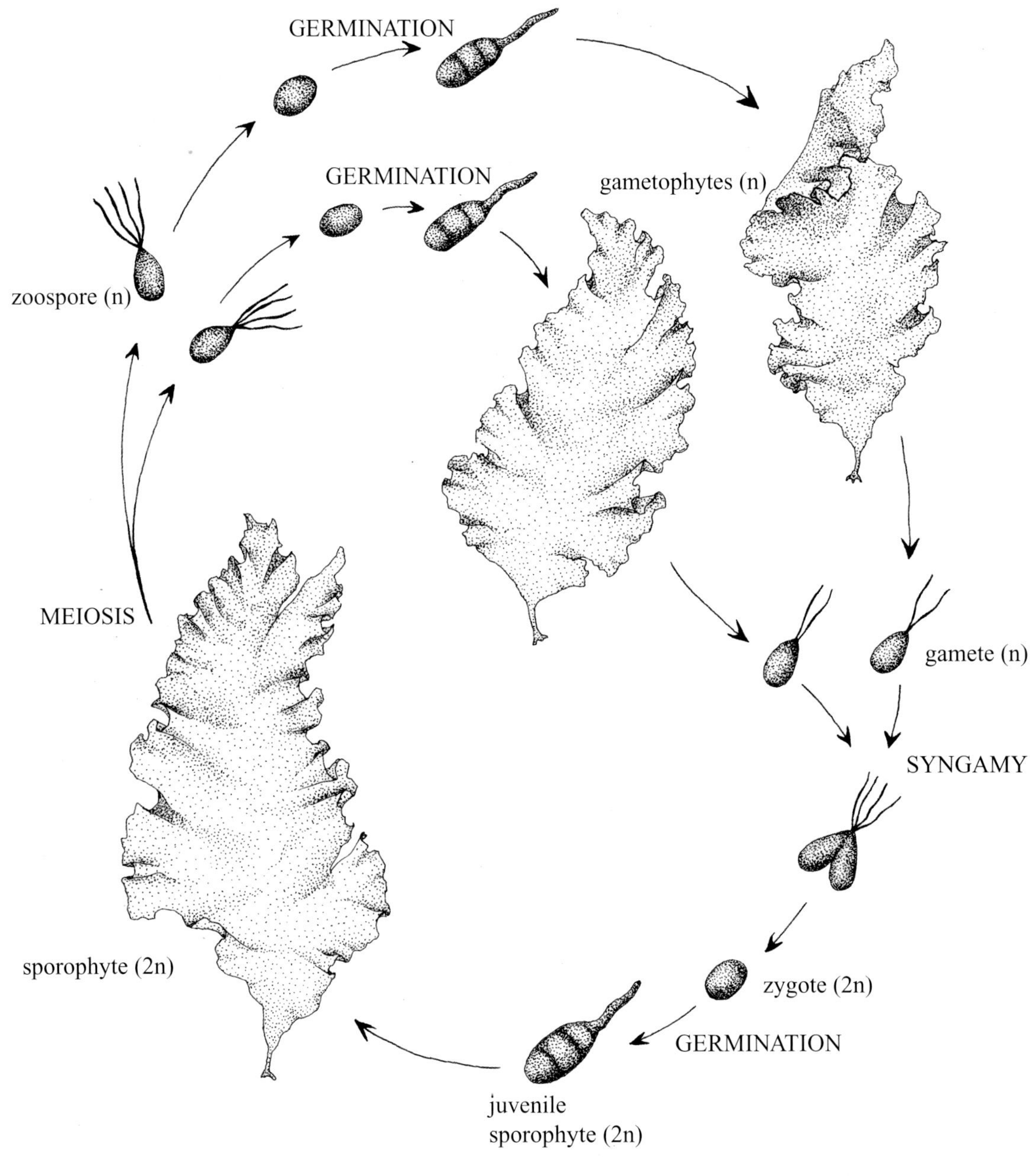

Figure 4. The life cycle of the green seaweed *Ulva* illustrating isomorphic alternation of generations. The diploid sporophyte is the site of meiosis and haploid zoospores are released. The spores germinate and grow into haploid gametophytes that are similar in size and shape to the sporophyte. Haploid, motile gametes are released, syngamy occurs, and the resulting zygote germinates and grows into a new sporophyte.

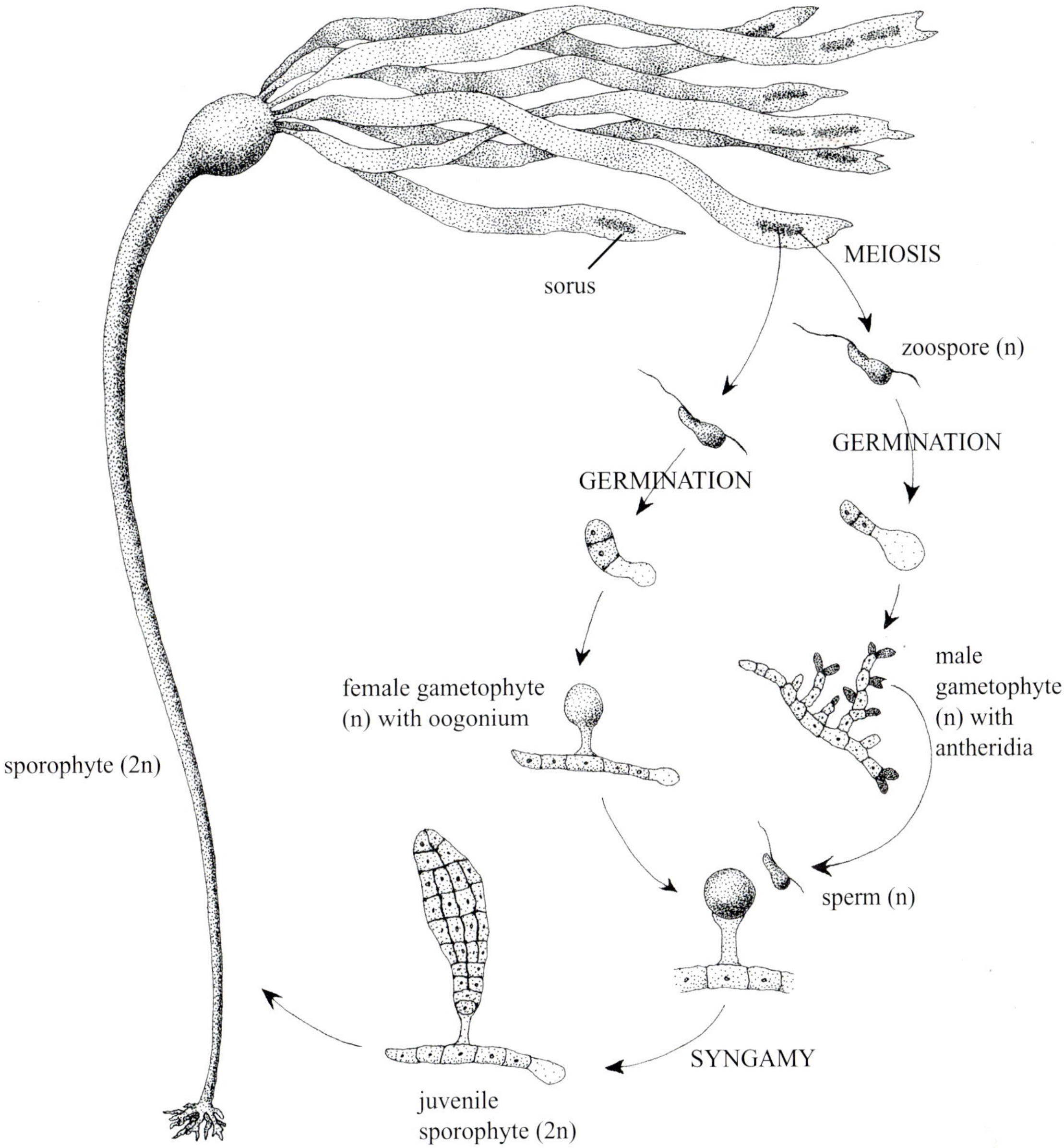

Figure 5. The life cycle of Bull Kelp, *Nereocystis,* demonstrating heteromorphic alternation of generations. A large, diploid sporophyte produces patches of reproductive structures called sori, where biflagellate zoospores are produced through meiosis. The haploid zoospores germinate and grow into microscopic gametophytes. Reproduction is oogamous: a nonmotile egg is fertilized by a small, biflagellate sperm cell. After syngamy, the diploid zygote germinates and grows into a sporophyte.

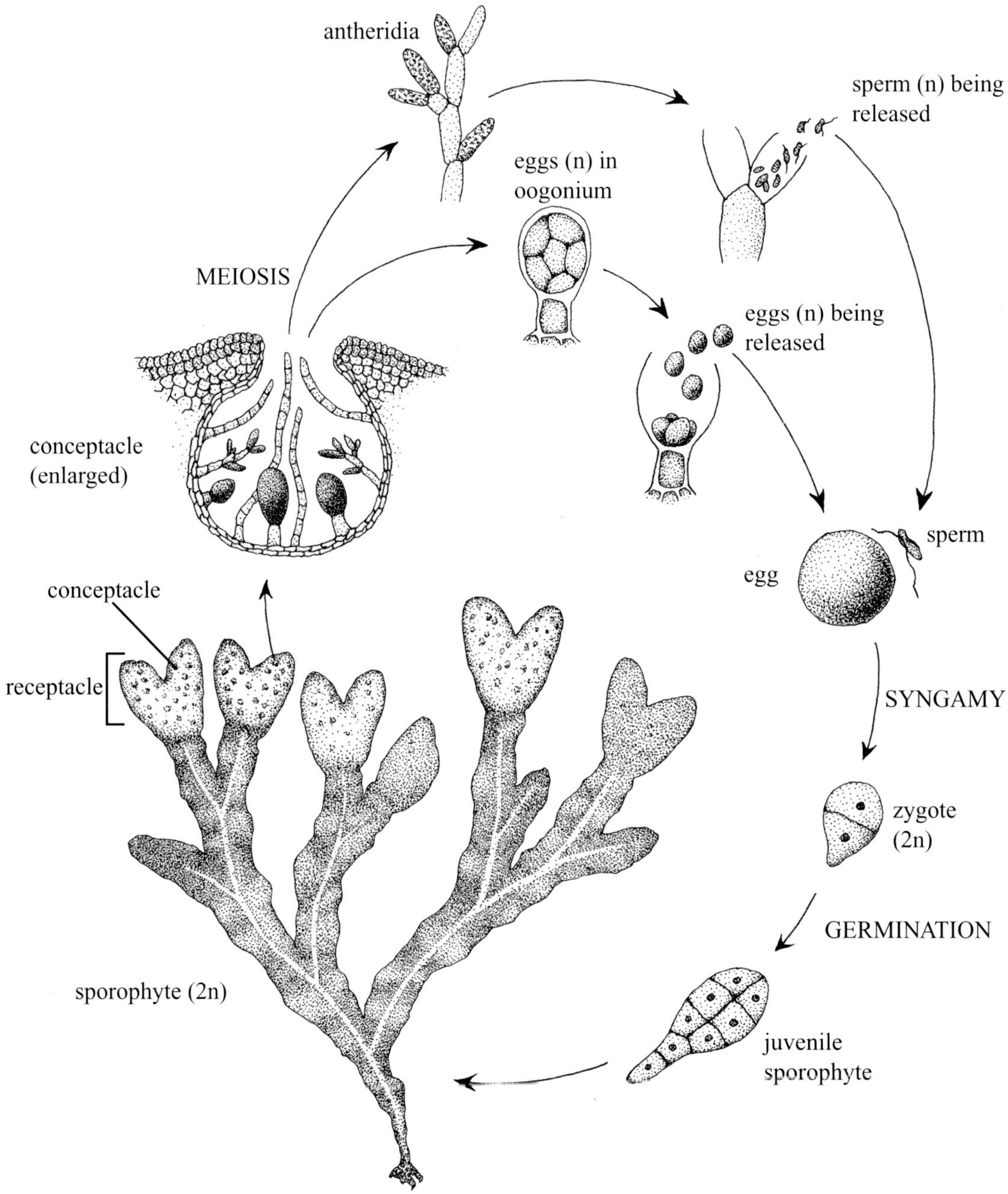

Figure 6. The life cycle of the brown seaweed *Fucus*. Meiotic division occurs in the diploid sporophyte and produces haploid gametes. The gametes are oogamous; a relatively large, nonmotile egg fuses with a smaller, motile sperm cell. A diploid zygote, the product of this fusion, germinates and grows into a new sporophyte. This type of life cycle is monomorphic ("single form"): the gametophyte is absent and there is no alternation of generations.

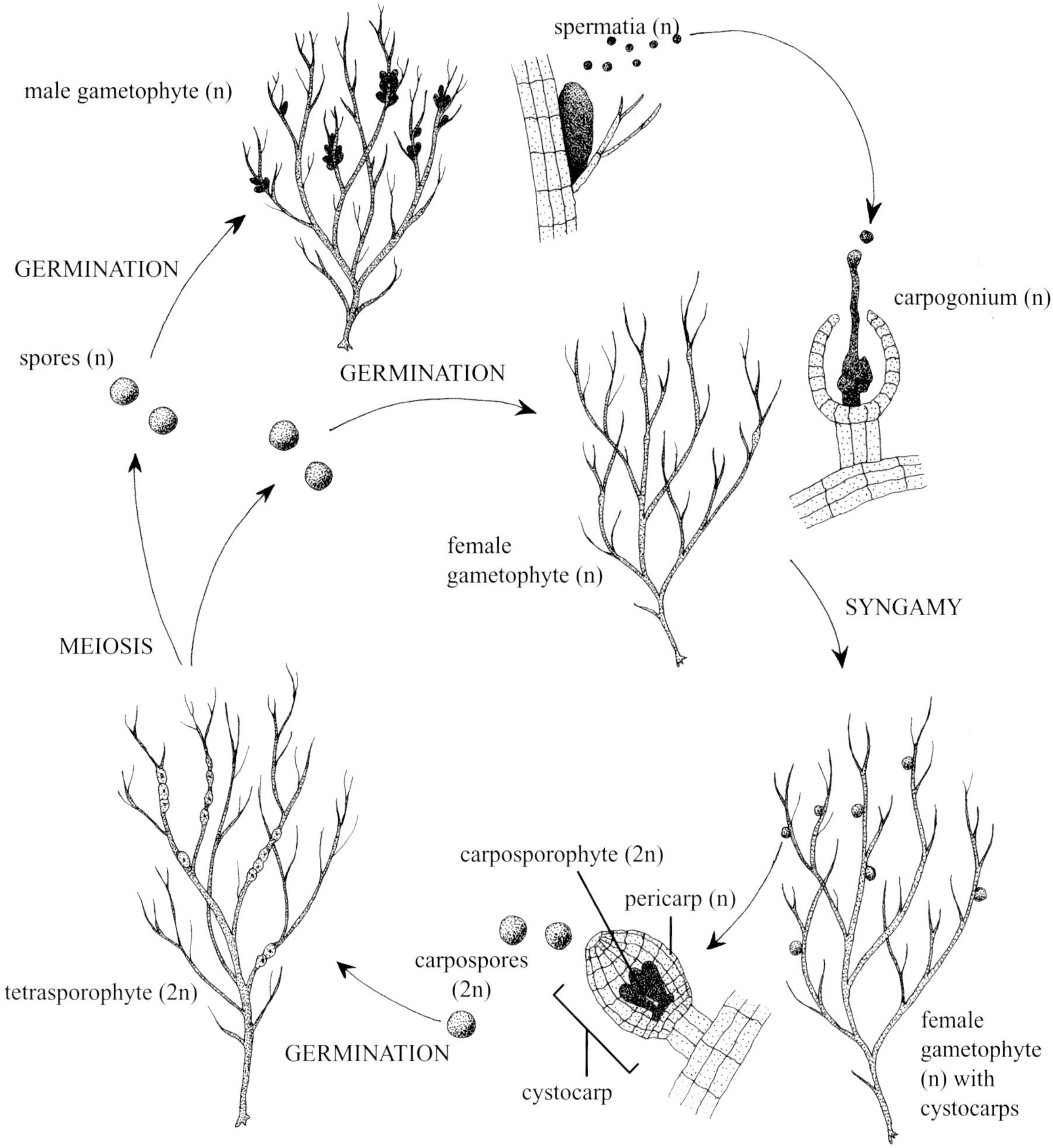

Figure 7. The life cycle of *Polysiphonia* illustrating the unique life history pattern of the red algae. A diploid tetrasporophyte produces nonmotile haploid spores via meiosis. The spores germinate and grow into gametophytes. Small, nonmotile spermatia drift through the water to the female gametophyte and fertilize the female reproductive cell, the carpogonium, while it is still attached to the gametophyte. A diploid, multicellular structure, the carposporophyte, is produced from syngamy. The carposporophyte continues to grow and live attached to the female gametophyte and is often surrounded by a protective structure, the pericarp, a product of the female gametophyte. The carposporophyte and the pericarp together are called the cystocarp. Spores are released from the carposporophyte and grow into a new tetrasporophyte.

often surrounded by a protective structure, the pericarp, that is a product of the female gametophyte. The composite structure of the carposporophyte and the pericarp is called a cystocarp.

The carposporophyte produces diploid mitospores, carpospores, that function as zygotes would in other life cycles by growing into independent diploid thalli in which meiosis occurs to produce haploid spores. As the haploid spores are produced in distinct sets of four, tetrads, they are called tetraspores and the thallus that produces them is called a tetrasporophyte.

Though all of the sexual reproductive cycles we've described are very common, in fact usually obligatory, many seaweeds can also increase their numbers by asexual cloning. This can occur when a fragment of the thallus breaks loose and grows into a new thallus. Asexual reproduction is more often accomplished through special unicells, spores, produced through mitotic division so that the spores have the same chromosome complement as the parent thallus that produced them. Mitospores in the green and brown algae are most often motile, red algal spores are always non-motile.

Uses for Seaweed

Sea Vegetables: Seaweed as Human Food

Seaweeds are an important source of food, especially in Asia; both cultivated and naturally growing seaweeds have been harvested in the Pacific Basin for hundreds of years. Algae are excellent sources of many vitamins including: A, B1, B2, B6, niacin, and C; they are also rich in iodine, potassium, iron, magnesium and calcium so it is not surprising that algae provide nutritional benefits. Seaweeds are incredibly versatile vegetables and can be eaten fresh, dried, pickled, or ground into powder. Roughly 20 species of algae are used in everyday cooking in Japan and at least 70 species are eaten in China. Some of the most important edible seaweeds are *Porphyra* (Nori), *Laminaria* (Kombu) *Palmaria* (Dulse), and *Undaria* and *Alaria* (Wakame). See the recipe section of the book for ideas on cooking with seaweeds.

Algae and Human Health

For centuries algae have been used in Asian countries to cure and prevent disease and currently there is a great deal of research looking for medicinal compounds in algae. The uses of algae in herbal medicine are diverse; it is suggested that, among other things, algae can help relieve constipation, gallstones, coughs, gout, high cholesterol, and high blood pressure. *Undaria*, a relative of *Alaria*, is said to enhance calcium uptake from the gut and is used to promote bone formation. Goiter (swelling of the thyroid gland) is commonly caused by inadequate intake of iodine and vitamin A; *Laminaria*, *Ulva*, *Porphyra* and *Sargassum* have traditionally been used to prevent goiter in mountainous regions of China, Peru, and Chile.

Digenea simplex, a red algae that grows in Japan, is prescribed in Asia as an effective herbal vermifuge to get rid of intestinal nematode worms (such as *Ascaris*). Its active ingredient, kainic acid, was isolated and identified in the early 1950's and is now used both as a vermifuge and in neurological research around the world. Some of the local seaweeds that are used against parasitic worms include: *Ulva*, *Codium*, *Corallina*, and *Sargassum*. Another interesting use of seaweed occurs in China where dried *Laminaria* stipes are used to absorb fluids and dilate the cervical canal during child birth.

Scientists have also discovered a class of anti-HIV chemicals in certain species of blue-green algae. The chemicals, called sulfolipids (or sulfonic acid-containing glycolipids) were previously known, but not known to be antiviral. This discovery resulted from a National Cancer Institute program to search systematically for new AIDS and cancer drugs that might be found in algae. In addition, recent research and anecdotal evidence on the anti-herpes action of extracts from red algae in the family Dumontiaceae has yielded much promise. Although tests have not been conducted on humans, researchers have also shown that extracts from *Spirulina* (a blue-green alga), *Sargassum*, *Codium*, and *Laminaria japonica* inhibit growth of oral cancer cells. Much research is needed to identify the specific algal components responsible for these health benefits and a lot of rigorous testing will be necessary to show their effectiveness, but algae show great promise as excellent sources for new drugs.

Seaweed as Fertilizers
Seaweeds have long been used as a source of fertilizer in many regions of the world. Seaweeds can be applied whole or chopped, and they can be used wet, dried, or composted. Some commercially available seaweed fertilizer is liquified or contains seaweed extracts. Fertilizer from seaweed is complex; it contains all the major and minor plant nutrients and has a high amount of potassium. Fertilizer from seaweed is a good complement to animal manure that contains about the same amount of nitrogen but is relatively high in phosphorus and only has about half the potassium as seaweed fertilizer. Seaweeds are also rich in trace elements and may contain hormones and growth regulators such as cytokinin. Another advantage of using seaweed fertilizer is that seaweeds are free of weeds and fungal spores.

Algal Extracts
Three important gelling agents used in the food, textile, cosmetic, and pharmaceutical industries come from seaweed: alginate, agar, and carrageenan. Alginate is formed from the alginc acid found in the cell walls of brown algae. Commercially, alginates are derived from large brown seaweeds generally growing in the colder water areas of the world; the compounds are used for thickening, suspending, stabilizing, emulsifying, or gel-forming. The main commercial sources for this gum are *Ascophyllum* (a close relative of *Fucus*), *Laminaria,* and *Macrocystis*. About half of the alginate produced is used for making ice cream and other dairy products like chocolate milk and yogurt.

Agar is best known to most of us as the culture growing medium used in petri dishes in school science labs. Dissolved in boiling water and cooled, agar becomes gelatinous and produces a firm, clear jelly that is rich in iodine and trace minerals. Agar is derived from a variety of red algae genera including: *Gelidium, Gracilaria, Pterocladia, Mastocarpus,* and *Ahnfeltia*. Cultivation and wild harvest of algae for agar occurs worldwide with large production areas in eastern Asia, Portugal, and California. Besides its use as a culture medium, agar is used in the pharmaceutical industry as a laxative and as an inert carrier for drug products where slow release of the drug is required.

Carrageenan is another gum extracted from red algae. Carrageenan is obtained from relatively few genera of red algae including: *Chondrus*, *Eucheuma,* and *Hypnea*. About half of the world's carrageenan is produced from algae cultivated in the Philippines and Indonesia, with another large production area in the north Atlantic. Chemically, both agar and carrageenan are polymers made up of the sugar, galactose, and they are components of algal cell walls. Both agar and carrageenan are used in the food industry as antidrying agents in breads and pastries and as a thickening agents in soups, sauces, salad dressings, jellies, and ice cream. All three of these extracts are used as a stabilizers for emulsions and are found in cosmetics, lotions, and shaving cream. These products also show up in photographic film, paint, shoe polish, and dental impression molds.

Seaweed Habitats

Tides

Tides, the alternating rise and fall of sea level in a day, are generated by the gravitational attraction between the earth, sun, and moon. Since the moon is much closer to the earth than the sun, it exerts a much stronger gravitational force on the planet and has more effect on the tides than the sun. When the earth faces the moon, the gravitational force affects the entire surface of the planet. The ocean, however, is the only portion that is flexible and large enough to be visibly affected by this force and a bulge of water forms on the side of the earth facing the moon. At the same time, centripetal force, created by the rotation of the earth, causes the ocean to bulge on the opposite side of the planet as well. High tide occurs where these two bulges of water are present.

In most places on earth, there are two high and two low tides each day; this tidal pattern is called semidiurnal. Along the west coast of North America, we generally have two high and two low tides of unequal height every day. This type of tidal pattern is called mixed semidiurnal.

The moon's rotation around the earth is the cause of the timing between high and low tides. The moon rotates around the earth at a rate of about 12 degrees a day, one rotation a month. This rotation is in the same direction as the earth's spin, so, by the time the earth has completed one rotation, the moon has shifted 12 degrees further and it takes an extra 50 minutes for the moon to be in the same location relative to a point on earth. For this reason, the tidal cycle is not 24 hours long, but 24 hours and 50 minutes long, thus high and low tides are about 50 minutes later each day.

Twice a month, during the full and new moon phases, the sun and moon are nearly in line and their gravitational forces combine. During these times there is an increase in the average range of the tides and these are known as spring tides. Alternatively, twice each month, the moon is in quarter phase and the sun and moon are at right angles to each other relative to the earth. During these times, we experience minimal tidal fluctuations, called neap tides. The best time to visit the shoreline looking for seaweeds, therefore, is at low tide during the spring tides.

Intertidal Zonation and Species Diversity

The section of shoreline periodically left uncovered through the rhythm of the rising and falling tides is known as the intertidal region. This region extends from the highest high water mark to the lowest region that is exposed on the most extreme tide. Although marine ecologists agree that the intertidal region can be divided into distinct patterns of zonation, there has been controversy over the years as to the proper method to define these intertidal zones. Some marine ecologists use tidal height to delineate the zones, while others base the zones on the presence or absence of particular species. When we describe algal habitats in this book, we simply use three intertidal regions: the upper, middle, and lower.

Below the lowest reaches of the intertidal zone is the subtidal zone. Many species of seaweeds, particularly the large kelps, inhabit the subtidal region. Subtidal seaweeds can only grow as deep as the sun light needed for photosynthesis will penetrate. This region is called the photic zone. In areas along the coast where the water is highly turbid, the photic zone is more shallow than in habitats where the water is clear.

Seaweed abundance and species diversity vary seasonally. The highest seaweed abundance and species diversity occur in late summer or early fall. Many kelp species, will be too large and cumbersome to collect then; late spring/early summer is the best time of year to find small undamaged specimens of kelp. There are, however, a few seaweed species that may only be found during the winter months.

Collecting Seaweeds

With the many uses for seaweed, it is not surprising that you might be tempted to start gathering your own. There are a few things to consider before harvesting seaweeds or aquatic plants.

When you are in the intertidal zone, it is important to keep an eye on the ocean at all times. Winter storms can produce large waves and it is best to avoid the tide-pooling during these times. Rogue or "sneaker" waves, however, can occur at any time. A little common sense and respect for the ocean will enable you to have safe and productive experiences in the intertidal region.

The laws for collecting seaweed vary with locality. Most west coast states require a fishing license and/or a permit and have specific laws governing the collection of seaweeds. Several species are protected and harvesting these species is illegal. Be sure to familiarize yourself with the local laws and regulations prior to collecting seaweeds or marine plants. If you choose to collect seaweed, for pressing or for eating, it is important to only take what you can use. Remember, an excellent alternative to foraging for algae is to order sea vegetables from companies that harvest seaweeds under controlled, sustainable conditions.

You should also be aware that your specimens will degrade quickly when removed from their natural habitat and that the best way to preserve seaweed is to dry them in an herbarium press. In order to keep your collections fresh until you have the time to press them, be sure your seaweeds are damp, but not wet. If you put your collection into plastic bags, lay the algae on paper towels to help separate specimens and absorb excess water. Be sure not to store members of the genus *Desmarestia* with your other specimens (see the *Desmarestia* section for more details). Keep seaweeds cool by storing them in the refrigerator (but don't forget about them, decomposing seaweed can be quite pungent!).

If you are gathering seaweeds for food, you can either use them while fresh or preserve them by drying or pickling.

Photographing Seaweeds

Photographing seaweeds can be both a challenging and rewarding experience. Seaweeds are photographically difficult because they are both highly reflective and "black holes" that absorb copious amounts of light. The reflective nature of seaweeds is exaggerated in the intertidal region because they are left wet as the tide recedes. Other than your tripod, the most important photographic accessories that will help you take successful seaweed photos in the intertidal region, are umbrellas and polarizing filters. Using umbrellas in conjunction with polarizing filters will reduce glare and hot spots caused by the sun's reflection on shiny seaweeds. We've received some funny stares as we carefully held umbrellas over wet seaweed while our cameras were sitting out in salt spray and rain, but using these tools will help produce more visually appealing photographs. Ultimately, however, the best conditions for photographing seaweeds in the intertidal region are overcast days which provide diffuse, even lighting.

The second factor to consider, and perhaps one of the biggest problems you will encounter when you set out to photograph marine algae, is their diverse size. Large kelps such as *Pelagophycus*, *Nereocystis* or *Macrocystis* may be 30 meters or more in length and present different challenges than the small foliose reds such as *Ptilota* and *Plocamium* that may be only a few centimeters in height. The following section outlines the gear and techniques we have found to be the most productive.

Large Kelps

The large kelps (> 2m) are predominately found subtidally and unless you have access to an extremely large aquarium (which most of us don't) photographing these subjects will be done in their natural environment. For the large kelps, we use both a Nikonos camera with a 15 mm lens and Nikon N90s cameras in Aquatica housings with 16 mm to 20 mm lenses. Both camera systems are equipped with Sea and Sea YS120 strobes. Obviously, good visibility is invaluable when photographing large subjects underwater.

Medium Sized Seaweeds

Medium sized seaweeds (2 m to about 0.5 m) can be found both intertidally and subtidally. When photographing in the intertidal region, a Nikon 35-70 mm zoom lens is one of our most used lenses. Subtidally, the versatile Nikonos V with 15 mm lens is our workhorse camera. At times, when a greater working distance is necessary, we use a housed system with either a 24 mm or 60 mm lens.

Small Seaweeds

Photographing small seaweeds (< 0.5 m) requires the use of more diverse techniques. Many of the small seaweeds are easily photographed in their natural habitat. Some of the smaller seaweeds, however, may need to be collected and then photographed so that their distinguishing characteristics can be more readily displayed. Whether photographing in the intertidal, an aquarium, or underwater using a housed camera system, the Nikon 60 mm micro and Nikon 105 mm micro are our lenses of choice. Both of these lenses are capable of 1:1 ratios (life-size) and allow you to capture the intricate detail of small seaweeds.

Taxonomy of Algae

Taxonomy is the branch of biology that deals with classifying and naming living things and grouping organisms based on related biological characteristics and genetic relationships. Biological classification is based on a hierarchical system developed by Carl Linnaeus in the eighteenth century. The system includes eight categories that become smaller and more inclusive; they are, from largest to smallest: Kingdom, Phylum, Class, Order, Family, Genus, Species.

At first glance, grouping and organizing organisms into this hierarchy seems logical and somewhat simple. In reality, the task is quite subjective and very problematic. Imagine a group of blocks of different sizes, shapes, and colors, made of several different materials. If you asked someone to organize these blocks, would the person organize them by color? Within each color would shape, or material, or size be the next level of classification? Or maybe material would be the highest level, then shape, and color and size would be last. It is quickly obvious that there is no correct way to organize the blocks-- lots of ways would be logical, it would just be a matter of opinion.

Until about the 1960s, all life was categorized into two kingdoms: Plants and Animals. But now another problematic aspect of classification arises: life is not as simple as colored blocks, and organisms often do not fit neatly into defined groups. By definition, it is agreed that a creature that runs around and consumes other organisms for food is an animal. Likewise, a rooted-organism that uses sunlight to create food is a plant. But how should we classify an organism that is mobile and ingests food particles but also uses sunlight to create food? One solution is to create a new kingdom. At present, the most common classification scheme is a five-kingdom system that includes: Kingdom Plantae, Kingdom Animalia, Kingdom Monera, Kingdom Fungi, and Kingdom Protoctista (Figure 8).

As scientists learn more about life cycles, development phases, and the genetics of organisms we are realizing that our five-kingdom classification system is an oversimplification and is as artificial as the old, two-kingdom system. Thus, the way we subdivide life into categories is going to continue to change and evolve. Many groups within various kingdoms are very distinct and probably deserve kingdom status; some biologists have proposed as many as 20 kingdoms to accommodate this diversity!

Where algae fit into this classification scheme will also continue to change. The group of organisms that we call algae is really a very diverse assemblage of unrelated organisms and, in many cases, we just don't know enough about them to list characters that would make logical classification characteristics. For example, water molds are usually classified as members of the kingdom Fungi, but recent evidence shows that they are probably more closely related to brown algae. On the other hand, green algae are much more closely related to plants than they are to other algae. For now, most algae are classified in the kingdom Protoctista. Within the kingdom Protoctista, seaweeds are divided into three main groups: green algae, brown algae, and red algae.

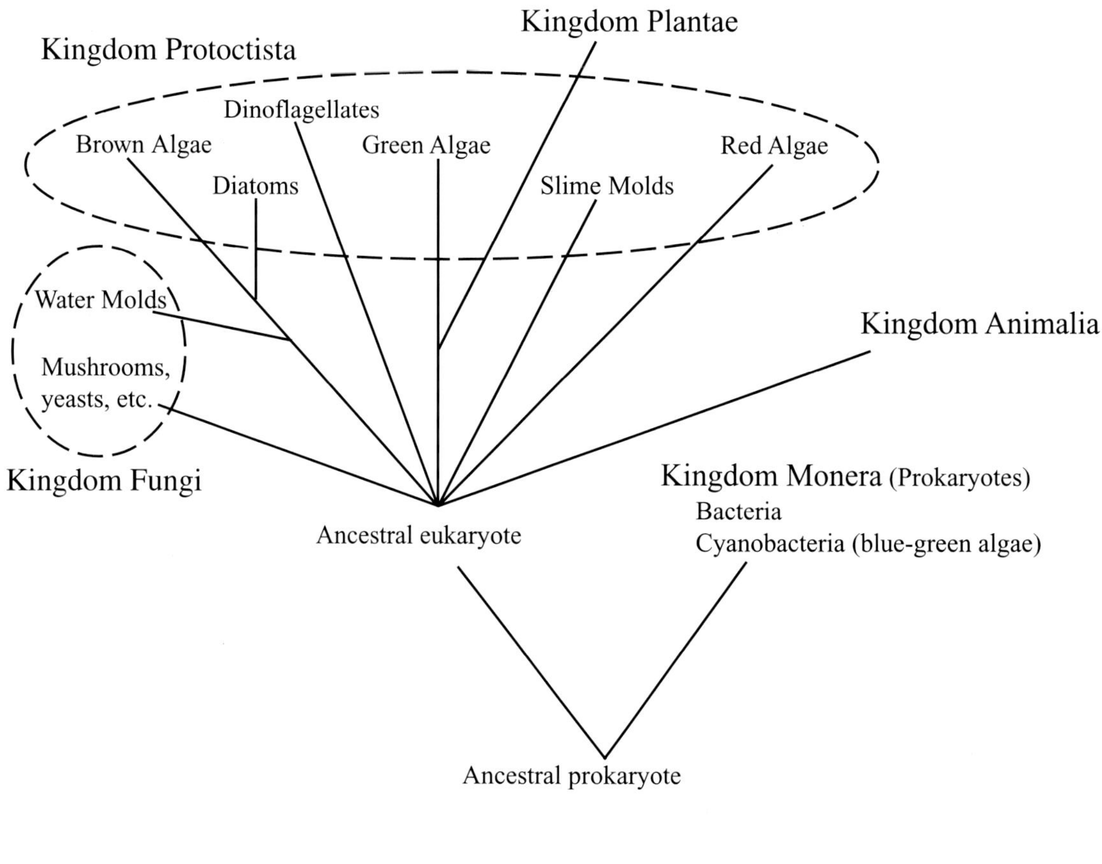

Figure 8. A simplified phylogenetic tree illustrating the five-kingdom scheme and the polyphyletic nature of algae and the rest of the Kingdom Protoctista.

Quick-Key

Quick-Key is designed to save you time by quickly sorting your specimens into small groups of likely species, before you resort to the text and illustrations. Start off by matching the color of your specimen with the appropriate table heading. Next, find the form category that best fits your specimen and cross reference it to your goegraphical area. Then, turn to the species descriptions and pictures.

Light to Dark Green in Color

	AK	SE AK	BC	WA	OR	N CA	S CA	Baja
FILAMENT								
Acrosiphonia coalita	■	■	■	■	■	■	■	
Acrosiphonia arcta	■	■	■	■	■			
Bryopsis corticulans			■	■	■	■	■	■
Chaetomorpha aerea			■	■	■	■	■	■
Cladophora columbiana		■	■	■	■	■	■	■
Ulothrix flacca	■	■	■	■	■	■	■	
Urospora penicilliformis	■	■	■	■	■	■	■	
CRUST								
Codium setchellii		■	■	■	■	■	■	■
TUBE								
Enteromorpha intestinalis	■	■	■	■	■	■	■	■
SPHERE								
Derbesia marina	■	■	■	■	■	■	■	■
DICHOTOMOUSLY BRANCHED								
Codium fragile		■	■	■	■	■	■	■
BLADE								
Kornmannia leptoderma	■	■	■	■	■	■	■	
Ulva spp.	■	■	■	■	■	■	■	■
GRASSLIKE APPEARANCE								
Phyllospadix spp.	■	■	■	■	■	■	■	■
Zostera spp.	■	■	■	■	■	■	■	■

AK = Bering Sea, Aleutian Islands, & Southcentral Alaska
SE AK = Southeast Alaska
BC = British Columbia
WA = Washington
OR = Oregon
N CA = Northern California
S CA = Southern California
Baja = Baja California

Yellowish-Brown to Olive-Green to Dark Brown

	AK	SE AK	BC	WA	OR	N CA	S CA	Baja
TUBE OR CYLINDRICAL AXIS								
Desmarestia aculeata	■	■	■	■	■			
Desmarestia viridis	■	■	■	■	■	■	■	■
Melanosiphon intestinalis	■	■	■	■	■	■	■	■
Scytosiphon dotyi					■	■	■	■
Scytosiphon lomentaria	■	■	■	■	■	■	■	■
GLOBULAR OR SAC								
Coilodesme californica	■	■	■	■	■	■	■	■
Colpomenia bullosa	■	■	■	■	■	■		
Colpomenia peregrina	■	■	■	■	■	■	■	
Leathesia difformis	■	■	■	■	■	■	■	■
Soranthera ulvoidea	■	■	■	■	■	■	■	■
DICHOTOMOUSLY BRANCHED								
Dictyota binghamiae			■	■	■	■	■	■
Fucus gardneri	■	■	■	■	■	■	■	
Hesperophycus californicus						■	■	■
Silvetia compressa						■	■	■
Pelvetiopsis limitata			■	■	■	■	■	
FLATTENED WITH PINNATE BRANCHING								
Desmarestia latifrons					■	■	■	■
Desmarestia ligulata	■	■	■	■	■	■	■	■
BUSHY, BRANCHING IN MANY DIRECTIONS								
Analipus japonicus	■	■	■	■	■	■	■	
Cystoseira osmundacea					■	■	■	■
Cystoseira geminata	■	■	■	■				
Egregia menziesii			■	■	■	■	■	■
Sargassum muticum		■	■	■	■	■	■	■
SMALL BLADE (<30cm)								
Petalonia fascia	■	■	■	■	■	■	■	■
Phaeostrophion irregulare	■	■	■	■	■	■	■	
Zonaria farlowii							■	■

AK = Bering Sea, Aleutian Islands, & Southcentral Alaska
SE AK = Southeast Alaska
BC = British Columbia
WA = Washington
OR = Oregon
N CA = Northern California
S CA = Southern California
Baja = Baja California

	AK	SE AK	BC	WA	OR	N CA	S CA	Baja
LARGE BLADE(S) (>30 cm), NO STIPE								
Hedophyllum sessile	■	■	■	■	■	■		
SINGLE STIPE, UNDIVIDED BLADE, WITHOUT MIDRIBS								
Laminaria bongardiana	■	■	■	■	■	■		
Laminaria ephemera		■	■	■	■	■		
Laminaria farlowii						■	■	■
Laminaria saccharina	■	■	■	■	■	■	■	
Laminaria setchellii	■	■	■	■	■	■	■	■
Laminaria sinclairii			■	■	■	■	■	
SINGLE STIPE, UNDIVIDED BLADE, WITH MIDRIBS								
Agarum fimbriatum		■	■	■	■	■	■	■
Alaria fistulosa	■	■						
Alaria marginata	▪	■	■	■	■	■	▪	
Alaria taeniata	■	■	■					
Costaria costata	■	■	■	■	■	■	■	
Cymathere triplicata	■	■	■	▪				
Pleurophycus gardneri	■	■	■	■	■	■	▪	
SINGLE OR MULTIPLE STIPES, MULTIPLE OR DIVIDED BLADES								
Dictyoneurum californicum			■	■	■	■	■	
Eisenia arborea			■	■	■	■	■	■
Lessoniopsis littoralis	▪	■	■	■	■	■		
Macrocystis spp.		■	■	■	■	■	■	■
Nereocystis luetkeana	■	■	■	■	■	■	▪	
Pelagophycus porra							▪	■
Postelsia palmaeformis			■	■	■	■	▪	
Pterygophora californica	▪	■	■	■	■	■	■	■

Pink to Bright Red to Brownish Red to Deep Purple or Black

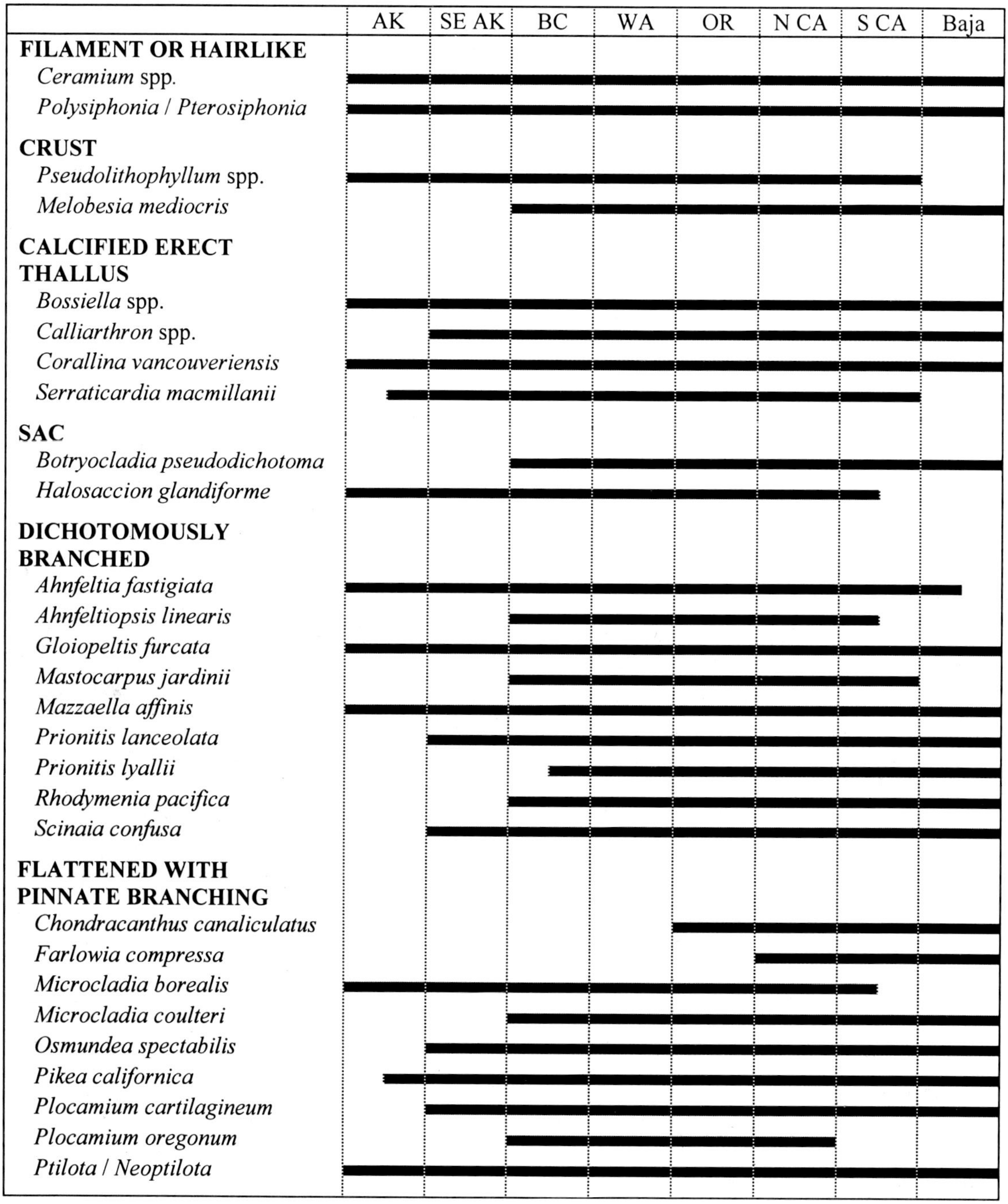

AK = Bering Sea, Aleutian Islands, & Southcentral Alaska
SE AK = Southeast Alaska
BC = British Columbia
WA = Washington
OR = Oregon
N CA = Northern California
S CA = Southern California
Baja = Baja California

	AK	SE AK	BC	WA	OR	N CA	S CA	Baja
BUSHY, BRANCHING IN MANY DIRECTIONS								
Callithamnion pikeanum	■	■	■	■	■	■	■	
Cryptosiphonia woodii	■	■	■	■	■	■	■	
Cumagloia andersonii	■	■	■	■	■	■	■	■
Endocladia muricata	■	■	■	■	■	■	■	■
Farlowia mollis	■	■	■	■	■	■	■	■
Gastroclonium subarticulatum		■	■	■	■	■	■	■
Gelidium / Pterocladia		■	■	■	■	■	■	■
Gracilaria / Gracilariopsis		■	■	■	■	■	■	■
Neorhodomela larix	■	■	■	■	■	■	■	■
Odonthalia floccosa	■	■	■	■	■	■	■	
Sarcodiotheca gaudichaudii			■	■	■	■	■	■
UNDIVIDED BLADE, NO VEINS								
Chondracanthus exasperatus		■	■	■	■	■	■	■
Chondracanthus corymbiferus			■	■	■	■	■	■
Constantinea simplex	■	■	■	■	■	■	■	
Constantinea rosa-marina	■	■						
Mazzaella splendens	■	■	■	■	■	■	■	■
Mazzaella oregona		■	■	■	■	■	■	
Opuntiella californica	■	■	■	■	■	■	■	■
Porphyra spp.	■	■	■	■	■	■	■	■
Smithora naiadum	■	■	■	■	■	■	■	■
DIVIDED BLADE, NO VEINS								
Callophyllis flabellulata	■	■	■	■	■	■	■	■
Callophyllis pinnata	■	■	■	■	■	■	■	■
Dilsea californica	■	■	■	■	■	■	■	
Fauchea laciniata	■	■	■	■	■	■	■	■
Fryeella gardneri			■	■	■	■	■	■
Halymenia spp.			■	■	■	■	■	■
Mastocarpus papillatus	■	■	■	■	■	■	■	■
Mazzaella parksii	■	■	■	■	■	■		
Palmaria callophylloides	■	■	■					
Palmaria hecatensis	■	■	■	■	■	■		
Palmaria mollis	■	■	■	■	■	■	■	
Sparlingia pertusa	■	■	■	■	■			
BLADE WITH VEINS								
Cryptopleura ruprechtiana		■	■	■	■	■	■	■
Delesseria decipiens	■	■	■	■	■	■	■	
Erythrophyllum delesserioides	■	■	■	■	■	■	■	
Hymenena flabelligera		■	■	■	■	■	■	
Membranoptera platyphylla	■	■	■	■	■	■	■	
Polyneura latissima	■	■	■	■	■	■	■	■

PHYLUM CHLOROPHYTA
Class Ulvophyceae

Order Cladophorales
Family Cladophoraceae

Chaetomorpha aerea (Dillwyn) Kützing

Description: A uniseriate, unbranched filament that is 5 to 30 cm long comprised of large cells (non fertile 125–400 μm diameter) that are visible if the filament is held up to light or can be felt by running your fingers along the strand. The filaments are grass-green and have white tips if spores or gametes have been released.

Pacific Coast Distribution: Southern British Columbia to Gulf of California, Mexico.

Previous name: *Conferva aerea* Dillwyn

Habitat: On rocks, near tide pools, mid to high intertidal.

Remarks: There is some taxonomic uncertainty about *Chaetomorpha aerea* and *C. linum* (Müller) Kützing. These have been designated as forms of the same species by some authors, as separate species by others. There are several other unbranched, filamentous species of green algae that could be confused with smaller celled species of *Chaetomorpha*, microscopic examination is needed for accurate identification.

Cladophora columbiana Collins — Green Tuft

Description: The alga is composed of uniseriate, branched filaments that form rounded, grass-green colored clumps that resemble the tufts of some mosses.

Pacific Coast Distribution: Southern southeastern Alaska to Baja California.

Previous names: *Cladophora hemisphaerica* Gardner; *Cladophora trichotoma* (C. Agardh) Kützing

Habitat: Abundant, on rocks, in the mid to high intertidal.

Remarks: This is the most abundant species of *Cladophora* on the Pacific Coast, but there are many other species that occur in both marine and fresh water.

Order Codiales
Family Bryopsidaceae

Bryopsis corticulans Setchell
Description: The thallus is 5–10 cm tall and consists of a central tubular axis, about 1 mm in diameter, with abundant, pinnately arranged secondary branches on the upper portion. The thallus is black-green out of the water, but appears bright green when submerged.
Pacific Coast Distribution: Ucluelet, British Columbia, to Chamela, Jalisco, Mexico.
Habitat: Frequent, on vertical sides of rocks, shells, wood and other algae, in the mid to low intertidal.

Derbesia marina (Lyngbye) Solier Sea Bottle ("Halicystis stage")
Description: The gametophyte is a small (2–15 mm diameter) green sphere with a rhizoidal base anchoring it to the substrate. The thalli are either male or female and bear patches of gametetangial cells; the patches are pale yellow-green on males and dark green on females. The sporophyte is an inconspicuous branched tube 0.5 mm in diameter and 1 cm long.
Pacific Coast Distribution: Aleutian Islands, Alaska, to Baja California.
Habitat: On rocks and encrusting coralline algae, in exposed habitats, in the low intertidal to subtidal (20 m).
Remarks: Before the life history of *Derbesia* was known, the sporophyte was thought to be a separate species *Halicystis ovalis* (Lyngbye) Areschoug.

Family Codiaceae

Codium fragile (Suringar) Hariot Dead Man's Fingers

Description: A dark green to blackish green thallus (10 to 40 cm tall) consisting of one to many dichotomously branched cylindrical segments. The branches arise from a broad, spongy basal disk.

Pacific Coast Distribution: Sitka, Alaska, to Bahia Asuncion, Baja California.

Habitat: Frequent to common, on rocky shores, in pools in the mid to low intertidal. Occasional to common, on rocks, subtidally.

Remarks: It is quite common to find *C. fragile* covered with red algal epiphytes. Although *C. fragile* appears to be native to the Pacific coast, a subspecies invaded the New England coast around 1957 and has spread rapidly since then. The plant was inadvertently brought to the area, probably as a fouling organism on ships from Europe. *Codium* has become a problem for the New England shellfish industry since the plant can foul nets and, when attached to shells, can cause shellfish to become buoyant and float away.

Codium setchellii Gardner

Description: The thallus is a thick, spongy crust, dark green to black in color and forms irregular cushions up to 25 cm broad.

Pacific Coast Distribution: Chichagof Island, Alaska, to Punta Baja, Baja California.

Habitat: Common on rocks, mid intertidal to upper subtidal.

Remarks: Several species of ascoglossan sea slugs occur on and eat *Codium* and *Bryopsis*. One of these sea slugs, *Elysia*, is able to retain functional chloroplasts after it sucks them out of the green algae; the chloroplasts carry on photosynthesis for several days inside the animal.

Order Ulotrichales
Family Acrosiphoniaceae

Acrosiphonia coalita (Ruprecht) Scagel, Garbary, Golden *et* Hawkes Green Rope

Description: The bright green to dark green thallus is composed of highly tangled branched filaments that form 10 to 20 cm long rope-like strands. Characteristic hook-shape branches (that can be seen with a hand lens or microscope) entangle adjacent filaments creating the frayed green rope appearance.

Pacific Coast Distribution: Aleutian Islands, Alaska, to San Luis Obispo, California.

Previous name: Spongomorpha coalita (Ruprecht) Collins; there is some debate about the taxonomy of this species and some phycologists call it *Acrosiphonia mertensii* (Ruprecht) Yendo

Habitat: Common to abundant, on rocks and in tidepools, in the mid to low intertidal.

Remarks: The rope-like plant is the gametophyte phase of *Acrosiphonia*. Cells in the filaments of the thallus form swimming gametes that are released, eventually fuse and grow into a tiny (less than 1 mm) sporophytes that live inside the thalli of red algae. When they were first seen inside the blades of red algae, these sporophytes were thought to be different species in the genus *Codiolum*.

Acrosiphonia arcta (Dillwyn) Gain

Description: A grass-green algae whose coarse, branched filaments form tufts up to 6 cm long. The tufts are matted together only near the rhizoidal base.

Pacific Coast Distribution: Bering Sea to southern Oregon.

Previous name: Spongomorpha arcta (Dillwyn) Kützing

Habitat: On rocks, mid to low intertidal.

Remarks: Some phycologists believe that several other species of *Acrosiphonia* (*A. saxatilis* and *A. spinescens*) should be grouped into *A. arcta*.

Urospora penicilliformis (Roth) Areschoug

Description: The thallus is composed of uniseriate, unbranched filaments that grow in patches creating a dark green, slippery, hairlike coating on rocks or wood.

Pacific Coast Distribution: Bering Sea, Alaska, to Channel Islands, California.

Previous names: Hormiscia penicilliformis (Roth) Fries; *Urospora mirabilis* Areschoug

Habitat: Common, on rocks and wood, in the high intertidal.

Remarks: The filament is the gametophyte stage in an alternation of heteromorphic generations and releases biflagellate swimming gametes. The sporophyte is microscopic and releases quadriflagellate zoospores.

Family Kornmanniaceae

Kornmannia leptoderma (Kjellman) Bliding

Description: The thallus is a cluster of fan-shaped, thin blades that are usually less than 5 cm long and grass-green in color.

Pacific Coast Distribution: Aleutian Islands, Alaska, to San Luis Obispo County, California.

Previous names: Monostroma zostericola Tilden; *Kornmannia zostericola* (Tilden) Bliding

Habitat: Epiphytic on *Zostera* and *Phyllospadix*, low intertidal and subtidal.

Remarks: The thalli start as a tiny, 3 mm in diameter, sacs that split and grow into the blades.

Family Ulotrichaceae

Ulothrix flacca (Dillwyn) Thuret

Description: The thallus is composed of uniseriate filaments that form bright green strings or entangled mats.

Pacific Coast Distribution: Beaufort Sea, Alaska, to Channel Islands, California.

Previous name: Ulothrix laetevirens (Hooker) Collins

Habitat: On rocks, common in northern portion of the range, rare to uncommon further south, mid to high intertidal.

Remarks: Each cell of *Ulothrix* has a single U-shaped chloroplast that can be seen with a compound microscope.

Family Ulvaceae

Ulva spp.

Sea Lettuce

Description: The grass-green thalli are easily recognized as thin, often transparent sheets. A microscopic examination of a blade cross section reveals that it is only 2 cell layers thick.
Pacific Coast Distribution: Alaska to Mexico.
Habitat: Common on rocks, algae, and wood, in the mid to low intertidal and upper subtidal.
Remarks: Their bright green color and abundance in the intertidal make the members of this genus some of the more conspicuous seaweeds. Species of *Ulva* can be difficult to distinguish and are usually differentiated by characteristics such as the shape of the blade, presence or absence of perforations and small "toothlike" projections on edge of the blade. *Ulva* is widely used for food. It can be dried, toasted or eaten fresh in salads and soups and other dishes. If you are harvesting it to eat, be sure to collect *Ulva* far from any potential pollution since many species are reported to be tolerant of organic and metal pollution.

Enteromorpha intestinalis (Linnaeus) Nees

Description: The thallus consists of a hollow tube up to 20 cm long and about 5 mm in diameter. Tubes are usually bright green but can become yellowish to white, especially near freshwater.
Pacific Coast Distribution: Chukchi Sea, Alaska, to Sonora, Mexico.
Habitat: Grows in dense mats on rocks, other algae, or free-floating in clumps. Abundant in the upper intertidal, especially in brackish water or near freshwater seeps.
Remarks: In cross section the walls of the tubular thallus are one cell layer thick. In another species, *Enteromorpha linza,* the thallus is tubular only at the base and the plant becomes very flattened and up to 10 cm wide.

PHYLUM PHAEOPHYTA
Class Phaeophyceae

ORDER ECTOCARPALES
Family Coilodesmaceae

Coilodesme californica (Ruprecht) Kjellman
Description: The thallus is a light olive-tan sac up to 30 cm long that grows epiphytically on *Cystoseira*. The base tapers to a short stipe, 1–3 mm long, with a small discoid holdfast.
Pacific Coast Distribution: Kenai Peninsula, Alaska, to Islas San Benito, Baja California.
Habitat: Epiphytic on *Cystoseira* in the low intertidal and subtidal.

Phaeostrophion irregulare Setchell *et* Gardner
Description: A crustose holdfast up to 20 cm in diameter gives rise to many irregularly tapering blades that are 6–25 cm tall and 1–4 cm wide. Blades are light brown to tan in color, thin and smooth.
Pacific Coast Distribution: Aleutian Islands and Yakutat Bay, AK, to Santa Barbara County, CA.
Habitat: On rocks, usually in sandy areas, mid to low intertidal.
Remarks: This species can be difficult to distinguish from *Petalonia* in the field. *Phaeostrophion*, however, has an extensive crustose holdfast and generally grows in intertidal habitats that are subject to sand burial.

Family Ralfsiaceae

Analipus japonicus (Harvey) Wynne Sea Fir

Description: This olive-green to tan alga has one or more erect thalli arising from a common crustose base. The erect portions have cylindrical axes up to 30 cm long and are covered on all sides by short (1–3 cm) branches.
Pacific Coast Distribution: St. Lawerence Island, Alaska, to Government Point, Santa Barbara County, California.
Previous names: *Heterochordaria abietina* (Ruprecht) Setchell *et* Gardner
Habitat: On rocks, in the upper to low intertidal.
Remarks: The crust is the overwintering stage and the erect portion begins to grow in March. The specific epithet of the previous scientific name, *Heterochordaria abietina*, refers to its resemblance to a sprig of fir of the genus *Abies*. *Analipus* is eaten in Japan where it is called 'matsumo.'

Family Leathesiaceae

Leathesia difformis (Linnaeus) Areschoug Sea Cauliflower

Description: This olive-green alga forms saccate or globular masses with many surface convolutions. The thallus, up to 10 cm in diameter but usually about 4–5 cm across, is an annual that appears in early spring to October.
Pacific Coast Distribution: Bering Sea to Baja California.
Previous names: *Leathesia nana* Setchell *et* Gardner; *L. amplissima* Setchell *et* Gardner
Habitat: Common in the mid to upper intertidal, on rocks or epiphytic on *Phyllospadix*, *Neorhodomela*, and *Odonthalia*. Can be very abundant in sheltered locations.
Remarks: When it is small, it is easy to confuse *Leathesia* with *Colpomenia bullosa*. See remarks under *C. bullosa* for ways to differentiate the two.

Family Punctariaceae

Melanosiphon intestinalis (Saunders) Wynne Dark Sea Tubes

Description: Dark brown thalli that are cylindrical, unbranched, and usually twisted. The thalli are hollow when mature but solid when young.

Pacific Coast Distribution: Aleutian Islands, Alaska, to San Mateo County, California; Punta Blanca, Baja California.

Previous names: *Myelophycus intestinalis* Saunders

Habitat: Occasional, on rocks, in mid to upper intertidal.

Soranthera ulvoidea Postels *et* Ruprecht

Description: A globose and hollow alga that grows epiphytically on *Neorhodomela* and *Odonthalia*. The olive-brown thallus, 3–5 cm in diameter, is covered with evenly distributed sori (containing spore producing structures) that give the alga a "warty" appearance.

Pacific Coast Distribution: Bering Sea and Aleutian Islands, Alaska, to Baja California.

Habitat: Common to abundant, epiphytic on *Neorhodomela* and *Odonthalia*, in the mid to low intertidal.

Remarks: These sacs are the sporophyte stage that produce swimming spores with flagella; this phase alternates with a microscopic, filamentous gametophyte that produces gametes.

The dark bumps on this species make it easy to distinguish it the field from other saccate or globose brown algal epiphytes like *Coilodesme* and *Leathesia*.

Family Scytosiphonaceae

Colpomenia bullosa (Saunders) Yamada
Brown Bag

Description: A globular or saccate yellowish-brown to olive-green thallus, 1–10 cm high, that grows from a broad, contorted base.
Pacific Coast Distribution: Alaska Peninsula to central California.
Previous names: *C. sinuosa* f. *deformans* Setchell *et* Gardner; *Scytosiphon bullosus* Saunders
Habitat: Common, on rocks and other algae, in the mid to low intertidal.
Remarks: When young, *Colpomenia* is easily confused with *Leathesia difformis*. At maturity the surface of *Colpomenia* is smooth in contrast to the surface of *Leathesia*, which is lobed and convoluted. These two saccate species can also be distinguished by pressing a portion of the thallus between your fingers. *Leathesia,* which is composed of filaments, will disintegrate while *Colpomenia* tends to tear. In cross section, *Colpomenia* is readily identified by being composed of large round cells in contrast to the internal filaments of *Leathesia*.

Colpomenia peregrina (Sauvageau) Hamel
Description: The globular thallus can grow to 12 cm in diameter. Plants grow singly and attach to substrate by a small base. The surface of the thallus is covered with clusters of small colorless hairs.
Pacific Coast Distribution: Aleutian Islands, Alaska, to La Jolla, California.
Previous names: *Colpomenia sinuosa* var. *peregrina* Sauvageau
Habitat: On rocks and other algae, in the low intertidal and upper subtidal.
Remarks: Although this species is native to the Pacific, it was first described in France where it is an invasive, non-native; it has also spread throughout Britain and seems to be living up to the root of it's trivial name, *peregrin*, which, in Latin, means wander or travel. *C. peregrina* can be differentiated from *C. bullosa* in the field by its more or less spherical shape and considerably larger size.

Petalonia fascia (O. F. Müller) Küntze
Sea Petals

Description: Several narrow and tapering smooth blades, 10 to 25 cm long, of greenish-brown to golden-brown color, arise from a tiny crustose holdfast.
Pacific Coast Distribution: Aleutian Islands, Alaska, to Baja California; Chile.
Previous names: *Ilea fascia* (Müller) Fries, *Phyllitis fascia* (Müller) Kutzing, *Petalonia debilis* (C. Agardh) Derbès *et* Solier, *Ralfsia californica* Setchell *et* Gardner
Habitat: Common, on rocks, in the mid to high intertidal, often found as an epiphyte on *Phyllospadix* and *Zostera*.
Remarks: *Petalonia* is often found growing with *Scytosiphon*.

Scytosiphon dotyi Wynne

Description: The cylindrical, unbranched thalli, up to 12 cm tall and 1–2 mm in diameter, are yellow brown in color usually without constrictions although they can be twisted. This alga often grows in dense clumps.
Pacific Coast Distribution: Coos Head, Oregon, to Baja California.
Habitat: Occasional to common, on rocks and in tidepools, in upper intertidal.

Scytosiphon lomentaria (Lyngbye) Link

Description: The cylindrical, olive-brown to yellow-brown unbranched thalli, 20–30 cm tall and 6 mm in diameter, are usually hollow with constrictions at regular intervals.
Pacific Coast Distribution: Bering Sea, Alaska, to Baja California.
Previous names: *Scytosiphon simplicissimus* (Clemente) Cermades
Habitat: Locally abundant, on rocks, in the low intertidal.
Remarks: In Japan, *Scytosiphon* is dried and eaten in soup.

ORDER DICTYOTALES

Family Dictyotaceae

Dictyota binghamiae J. Agardh

Description: The thallus is tan to golden brown, 15 to 25 cm tall, and composed of smooth, thin, blades that lack a midrib. Branching is irregularly dichotomous (Y-shaped) and the holdfast is composed of creeping horizontal branches.

Pacific Coast Distribution: Queen Charlotte Islands, British Columbia, to San Juan Islands, Washington; sporadically from Washington to Monterey; common from Monterey to Baja California.

Habitat: On rocks, subtidal.

Remarks: In southern California, a similar species, *Dictyopteris undulata,* often grows in the same habitat but can be differentiated by a the presence of a distinct midrib.

Zonaria farlowii Setchell *et* Gardner

Description: The fan-shaped thallus, 8 to 15 cm tall, has concentric bands or stripes. The blades have no midrib and divide into many lobes creating a bushy thallus. In older plants the base of the blades can become stipe-like if the marginal tissue erodes.

Pacific Coast Distribution: Santa Barbara, California, to Isla Magdalena, Baja California.

Habitat: Common, on rocks, in the low intertidal and subtidal.

Remarks: This perennial alga can survive many months buried under sand.

Order Desmarestiales
Family Desmarestiaceae

Desmarestia ligulata (Lightfoot) Lamouroux
Acid Kelp

Description: The thallus, up to 1.5 m in length, is composed of a flattened main axis, 1–3 cm wide, with opposite branches. There are conspicuous mid-veins and finer, opposite-branched veins on both the main axis and branches. The sides of the main axis and all levels of branches may have marginal spines or hairs.

Pacific Coast Distribution: Unalaska (Aleutian Islands) and Kodiak Island, Alaska, to Baja California.

Previous names: *Desmarestia herbacea* Setchell *et* Gardner

Habitat: Frequently found on rocks in the low intertidal and subtidal.

Remarks: The common name comes from the fact that this, and many other members of Desmarestiaceae, secrete sulfuric acid. Take care not to mix *Desmarestia* with other specimens when collecting. A very similar species, *D. munda*, can be differentiated by wide branches and main axis (5–10 cm) and its habitat, which is strictly subtidal. Some phycologists believe these are the same species.

Desmarestia aculeata (Linnaeus) Lamouroux
Witch's Hair, Landlady's Wig

Description: This dark to light brown, wiry thallus may grow up to 2 m long. A discoid holdfast gives rise to the main axis which is cylindrical, about 2 mm in diameter, and bears profuse branches that can have alternate or irregular branching patterns.

Pacific Coast Distribution: Bering Sea, Alaska, to Coos Bay, Oregon.

Previous names: *Desmarestia intermedia* Postels *et* Ruprecht (Some phycologists consider this a separate species.)

Habitat: Common in pools and on rock, in the low intertidal and subtidal.

Remarks: Commonly found washed up on beaches after storms.

Desmarestia latifrons Kützing

Description: A discoid holdfast gives rise to a straight, cylindrical, though often somewhat flattened, main axis that grows up to 2 m in length. The branching pattern from the axis is irregular or alternate. The branches are flattened and leaflike with an inconspicuous midrib. Tiny hairs may be seen on the edges of the branches.

Pacific Coast Distribution: Arch Cape, Oregon, to Baja California.

Habitat: Common on rocks exposed to strong surf and in heavily sanded areas, in low intertidal and subtidal.

Remarks: Peters *et* al. 1997 examined rDNA from *Desmarestia* and their data suggest that *D. latifrons* is very closely related to *D. aculeata*; further study may show these to be northern and southern forms of the same species.

Desmarestia viridis (O. F. Müller) Lamouroux Stringy Acid Kelp

Description: The thallus is lax and fragile, up to 50 cm tall, with a cylindrical stipe that arises from discoid holdfast. The majority of the prolific branches arise from the central axis as opposite pairs.

Pacific Coast Distribution: Aleutian Islands, Alaska, to Baja; Chile; Argentina; Antarctic and North Atlantic.

Previous names: *Desmarestia media* var. *tenuis*

Habitat: Relatively uncommon, extreme low intertidal and subtidal.

Order Laminariales
Family Alariaceae

Alaria fistulosa Postels *et* Ruprecht
Description: The thallus consists of a well developed holdfast of haptera, a short stipe (20–30 cm) and a single, large, vegetative blade (up to 30 m) with a distinct midrib. The midrib of *A. fistulosa* is large (2–3 cm wide), hollow, and bears gas-filled chambers that distinguishes it from other species of *Alaria*.
Pacific Coast Distribution: Widespread throughout the Aleutian Islands, the Gulf of Alaska, and southeastern Alaska.
Habitat: On rocks, especially along shores exposed to some surf or current, in the low intertidal and subtidal.
Remarks: The gas chambers in the midribs lift the large vegetative blades toward the surface. *A. fistulosa* often forms large kelp beds, frequently associated with *Nereocystis luetkeana*.

Alaria marginata Postels *et* Ruprecht
Ribbon Kelp
Description: The small, well developed holdfast composed of haptera gives rise to a single cylindrical stipe about 5 mm in diameter and 7 cm long. A conspicuous midrib extends from the stipe through a single blade that can reach 3 m in length. In mature plants, two opposing rows of sporophylls are present along the stipe just below the base of the main blade.
Pacific Coast Distribution: Kodiak Island, Alaska, to Point Conception, California.
Previous name: *Alaria valida* Kjellman *et* Setchell
Habitat: On rocks, in the low intertidal and upper subtidal.
Remarks: There are at least seven species of *Alaria* on the Pacific coast, many of them are common in the north Pacific, but the range of *A. marginata* extends the furthest south.

Ribbon Kelp is a popular edible; both the blade and midrib can be eaten fresh, dried, or cooked. A somewhat similar species that is not native to this coast, *Undaria pinnatifida*, is widely used for food in Japan where it is called *wakame*. North American wakame is a common culinary term for *A. marginata*.

Alaria taeniata Kjellman
Description: The short cylindrical stipe becomes flattened where the sporophylls attach. Like other species of *Alaria*, the stipe leads to a single blade with a distinct midrib. The blade of *A. taeniata* is usually covered with tufts of tiny hairs, a characteristic rarely conspicuous in other species of *Alaria*.
Pacific Coast Distribution: Bering Sea and Aleutian Islands, Alaska, to southern British Columbia.
Previous names: *Alaria lanceolata* Kjellman
Habitat: On rocks, in the upper subtidal and low intertidal, in relatively sheltered habitats.
Remarks: *Alaria taeniata* is reportedly the most common species of intertidal *Alaria* in the Gulf of Alaska.

Egregia menziesii subtidally in Southern California.

E. menziesii in the Oregon intertidal.

Egregia menziesii (Turner) Areschoug
Feather Boa Kelp
Description: The thallus consists of a large holdfast and a stipe that divides irregularly into several branches that may be up to 10 m long. The branches bear a fringe of numerous small blades (up to 8 cm long) and oblong floats. The shape of the branches changes along their length: they are cylindrical near the holdfast, then become flattened and strap-like, finally there is a transition zone and the branches are blade-like in the upper portions.
Pacific Coast Distribution: Queen Charlotte Islands, British Columbia, to Baja California.
Habitat: Common on rocks in moderately exposed areas in the low intertidal and subtidal (to 6 m).

Eisenia arborea Areschoug

Southern Sea Palm

Description: A rigid woody stipe, up to 1 m long, arises from a large holdfast of haptera. The stipe is cylindrical at the base and becomes flattened; it then forks into two short branches that support numerous blades and sporophylls. The blades are covered with many irregular furrows and have dentate margins.

Pacific Coast Distribution: Graham Island, British Columbia, to Baja California.

Habitat: Common, on rock, low intertidal and subtidal, often forming dense groves.

Remarks: Young *Eisenia* look very different from adults; the juvenile specimens consist of a short stipe and a single broad blade with pointed toothlike projections on the margin.

Eisenia is commonly called Southern Sea Palm because it was once thought to occur only south of Point Conception.

Pterygophora californica Ruprecht

Description: A holdfast of haptera gives rise to a woody stipe 1–2 m long that is cylindrical at the base and becomes flattened. The stipe terminates in a single oblong blade 6–10 cm wide and up to 60 cm long. Sporophylls similar in shape and size to the terminal blade arise in two rows along each edge of the stipe.

Pacific Coast Distribution: Cook Inlet, Alaska, to Baja California.

Habitat: Common, on rocks, in the subtidal (2–20 m) where it can form dense stands.

Remarks: *Pterygophora* is a long-lived perennial kelp; the blade and sporophylls usually disintegrate during the winter and new ones are produced in the spring. Concentric rings can be seen in a cross section of the stipe that may represent growth rings.

Family Laminariaceae

Agarum fimbriatum Harvey — Fringed Sieve Kelp

Description: A single, bullate blade grows from a somewhat flattened stipe. A midrib (approximately 1–3 cm wide) arises as a continuation of stipe. There are numerous branched fimbriae along margin of stipe. The holdfast is composed of numerous haptera.

Pacific Coast Distribution: Duke Island, Alaska, to Puget Sound, Washington; Santa Cruz Island and San Diego, California; Islas Todos Santos, Baja California.

Habitat: On rocks, wood or other algae, subtidal .

Remarks: Differentiated from a very similar species, *Agarum clathratum*, by the bullate blade, flattened stipe, and numerous fimbriae along margin of stipe.

Costaria costata (C. Agardh) Saunders — Five-Ribbed Kelp

Description: A single, cylindrical stipe about 5 cm long that arises from a holdfast composed of haptera, terminates in a single, undivided blade. Five longitudinal ribs run the length of the blade. In between the ribs, the blade is often bullate. The color is dark to yellowish brown. Mature plants are 1.5 to 2 meters tall.

Pacific Coast Distribution: Aleutian Islands, Alaska, to Monterey County, California and San Miguel Island, California.

Habitat: Common, in pools and on rocks, in the low intertidal and shallow subtidal.

Cymathere triplicata (Postels *et* Ruprecht) J. Agardh Three-Ribbed Kelp

Description: A discoid holdfast gives rise to a stipe, 15 to 25 mm long, that supports a single undivided blade up to 4 m long and 20 cm wide. Three ridges with shallow troughs between them extend the length of the blade. The plant is yellowish brown. In favorable conditions the blade gets thick and leathery, and the longitudinal troughs can be 2 cm deep.

Pacific Coast Distribution: Bering Sea and the Aleutian Islands, Alaska, to northern Washington.

Previous names: *Laminaria triplicata* Postels *et* Ruprecht

Habitat: On rocks, low intertidal to upper subtidal.

Remarks: *Cymathere* is an annual kelp, by July most plants begin to fall apart and become tattered.

Hedophyllum sessile (C. Agardh) Setchell Sea Cabbage

Description: One of the few kelps without a stipe. The holdfast, consisting of branched haptera, arises directly from the blade. The blade which can be smooth or bullate, is thick and leathery but tears easily and, as the plant ages, the blade rips into strips.

Pacific Coast Distribution: Aleutian Islands, Alaska, to Monterey County, California.

Previous names: *Hedophyllum subsessile* (Areschoug) Setchell

Habitat: Common, on rocks, mid to low intertidal.

Remarks: *Hedophyllum* exhibits different growth forms depending on the amount of wave action where it is growing; in sheltered habitats it can be broad and highly corrugated whereas in exposed areas the blade will be smooth and have many longitudinal splits.

Laminaria bongardiana Postels *et* Ruprecht

Description: This kelp has a small holdfast of haptera and a single stipe of variable length (1 to 60 cm). The stipe supports a thick, dark brown blade, up to 2 m long, that is either smooth or somewhat corrugated and is usually split longitudinally 2 or 3 times. The base of the blade can be broad and rounded or heart shaped.
Pacific Coast Distribution: Bering Sea, Alaska, to San Francisco Bay, California.
Previous names: There is uncertainty about the taxonomy of this species and it is called *Laminaria groenlandica* by some phycologists.
Habitat: On rocks, low intertidal (especially in wave exposed areas) and subtidal.
Remarks: Species of *Laminaria* demonstrate wide morphological variation and environmental factors can dramatically change characteristics such as blade shape and thickness, the presence and absence of bullae, and splits in blade. Recent molecular studies have shown that defining species by these characteristics alone is probably not valid.

Laminaria farlowii Setchell Oarweed

Description: A branched holdfast leads to a single short (4–7 cm) stipe that supports a large, unbranched blade. The blade can reach up to 5 m in length and is covered by small wrinkled depressions (bullae).
Pacific Coast Distribution: Santa Cruz, California, to Baja California.
Habitat: Subtidal (down to 150 feet in southern California).
Remarks: This species is occasionally reported from British Columbia and Washington.

Laminaria saccharina (Linnaeus) Lamouroux Sugar Kelp

Description: This olive- to light-brown kelp is attached to the substrate with a holdfast composed of many, branched haptera. A single stipe (10–20 cm long) supports the blade (up to 3 m long) which often has 2 rows of corrugations (bullae) and can be longitudinally split.

Pacific Coast Distribution: Aleutian Islands, Alaska, to southern California.

Habitat: Abundant, (but rare in the southern portion of the range), on rocks, low intertidal and subtidal.

Remarks: The external morphology of *Laminaria saccharina* and *L. bongardiana* is variable and they can be difficult to distinguish. Internally, mucilage ducts are absent in the stipe of *L. saccharina* and they are present in *L. bongardiana.*

Laminaria setchellii Silva

Description: The thallus is up to 1 m tall and consists of a holdfast of haptera, a single stipe (up to 50 cm long), and a smooth blade that is deeply split into many uniform segments.

Pacific Coast Distribution: Attu Island (Aleutian Islands), Alaska, to Baja California.

Previous names: *Laminaria andersonii* Setchell; often misidentified as *L. dentigera.*

Habitat: Common, on rocks, low intertidal and subtidal.

Remarks: A similar species, *Laminaria dentigera*, is found in Alaska and northern British Columbia.

Laminaria sinclairii (Harvey *ex* Hooker f. *et* Harvey) Farlow, Anderson *et* Eaton

Description: The holdfast is composed of an extensive mass of prostrate, branched haptera. Numerous cylindrical stipes arise from the holdfast system, each bearing a single, smooth blade that is about 5 cm wide.

Pacific Coast Distribution: Hope Island, British Columbia, to Los Angeles County, California.

Habitat: Abundant, on rocks, in the low intertidal, often associated with sand.

Remarks: Another species, *Laminaria longipes*, shares the characteristic of a numerous stipes arising from one holdfast; it is found from the Bering Sea to southeast Alaska and northern Washington. The two species are differentiated by the presence mucilage ducts in the stipe of *L. sinclairii,* which are absent in *L. longipes*.

Laminaria ephemera Setchell

Description: An annual kelp, up to 2 m tall, that attaches to the substrate with a discoid holdfast. The holdfast supports a stipe and a single, smooth blade that is usually less than 10 cm wide.

Pacific Coast Distribution: Sitka, Alaska, to Monterey County, California.

Habitat: Common, on rocks, in the low intertidal and upper subtidal.

Remarks: *Laminaria yezoensis*, which also has disc-like holdfast, is a large, tough kelp with a blade over 20 cm wide and is found in Alaska.

Pleurophycus gardneri Setchell *et* Saunders *ex* Tilden

Description: The dark brown thallus has a cylindrical stipe, about 60 cm long and 1 cm in diameter, and a single blade up to 1 m long and 40 cm wide. The elastic blade has a distinct midrib up to 4 cm wide; delicate puckers occur in the blade along both sides of the midrib.

Pacific Coast Distribution: Aleutian Islands, Alaska, to San Luis Obispo County, California.

Habitat: On rocks, low intertidal to upper subtidal; rare in southern portion of range where it is found in the subtidal at depths exceeding 50 ft.

Remarks: Pleurophycus is a perennial deciduous kelp; the stipe and holdfast are perennial but the blade drops off in the fall.

Family Lessoniaceae

Dictyoneurum californicum Ruprecht

Description: A dichotomously branched, flattened and prostrate stipe that is attached to the substrate with forked haptera. The stipe supports multiple blades which are up to 1 m long and 4–8 cm wide and covered with a rectangular pattern of ridges. Edges of the blades often have toothlike projections.

Pacific Coast Distribution: Port Renfrew, Vancouver Island, British Columbia, to Channel Islands, California.

Habitat: Common, on rocks in exposed sites, in low intertidal and subtidal (to 10m).

Remarks: *Dictyoneurum* grows in dense patches and it is common to find pieces of the blade of this alga, with its characteristic reticulate ridge pattern, washed up in drift on the beach. A similar, but less common species, *Dictyoneuropsis reticulata*, is differentiated by the presence of a conspicuous midrib on the blade.

Lessoniopsis littoralis (Tilden) Reinke

Description: A very distinct feature of this kelp is the huge holdfast which can be 20 cm in diameter and 40 cm long. The holdfast is composed of many branched haptera that are very tough and woody. A mass of Y-shaped branched stipes arise from the holdfast, each bearing a single, narrow unbranched blade. Blades grow up to 1 m long and are 2–3 cm wide with a flattened midrib. Old plants may have 500 or more blades.

Pacific Coast Distribution: Kodiak Island, Alaska, to Monterey County, California.

Habitat: Common, on rocks in low intertidal, in areas exposed to high surf action.

Remarks: *Lessoniopsis* is a perennial and can live for many years.

Postelsia palmaeformis Ruprecht Sea Palm

Description: An annual kelp, up to 60 cm tall, that, during the summer, forms dense stands in rocky habitats pounded by surf. A small, stout holdfast supports a cylindrical stipe that is topped with a cluster of narrow blades. The blades are covered with longitudinal grooves that contain sporangia.

Pacific Coast Distribution: Hope Island, British Columbia, to San Luis Obispo County, California.

Habitat: Locally abundant, in exposed rocky habitats, mid to low intertidal.

Remarks: Considered a delicacy, the stipe and the blades can be pickled, steamed, or eaten fresh. In California, due to its popularity as an edible, *Postelsia* is protected and unlicensed harvesting is prohibited. *Postelsia* is named after the Russian naturalist, Postels.

Diver in a *Macrocystis pyrifera* kelp forest..

Close-up of *Macrocystis* fronds.

Macrocystis spp. (Linnaeus) Agardh Giant Kelp

Description: A huge kelp that grows up to 50 m and is one of largest algae. Numerous stipes, often 4 or 5 times dichotomously divided near the base, arise from a holdfast. Blades occur at regular intervals along the stipe. Mature blades are wrinkled in an irregular pattern, have toothlike projections along the edges and are fastened to the stipe by a basal pneumatocyst.

Habitat: Frequent, grows in dense kelp forests, on rocky shores, low intertidal and subtidal.

Pacific Coast Distribution: Alaska to Baja California.

Remarks: There are two species of *Macrocystis* along our coast and they are differentiated by their holdfast morphology. *M. pyrifera* has a large, conical holdfast that, in older plants, may be 1 meter tall, and is composed of branched haptera. *M. integrifolia* has a flattened, prostrate holdfast with numerous haptera that arise from its margin. *M. integrifolia* is a northern species; it is found from Alaska to Monterey, California, in the low intertidal and subtidal. The main distribution of *M. pyrifera* is Monterey, California, to Baja California, but there are also suspected populations in British Columbia and we have observed it growing near Sitka, Alaska. *M. pyrifera* is found subtidally, generally on rocky substrate, in depths ranging from 6 to 30 meters (depending on water clarity).

Macrocystis forests create an important three dimensional habitat for fish, invertebrates, and marine mammals such as sea otters. The structure of kelp forests is dynamic and can change depending on factors such as storms, water temperature and herbivory. In southern California, *Macrocystis* is extensively harvested, mainly for algin.

Nereocystis luetkeana (Mertens) Postels *et* Ruprecht Bull Kelp, Bullwhip Kelp

Description: *Nereocystis*, a large kelp that can reach lengths of 30 m, attaches to rocks with a holdfast that is composed of haptera. The holdfast supports a single, narrow, cylindrical, and whip-like stipe; along it length, the stipe gradually increases in diameter, becomes hollow, and terminates in a pneumatocyst (up to 15 cm diameter). From the top of the float two clusters of blades arise. Each cluster may contain up to 50 blades, which are up to 4.5 m long and 15 cm wide.

Pacific Coast Distribution: Aleutians Islands, Alaska, to San Luis Obispo County, California.

Habitat: Abundant, on subtidal rocks, forming dense kelp beds.

Remarks: *Nereocystis* is able to grow very fast, 15 cm per day in Washington, and thus reaches its mature size in just one season. Bull Kelp is edible, the stipe can be made into pickles or salsa and the blades can be dried and eaten like chips or added to soup and other food.

Pelagophycus porra (Leman) Setchell Elk Kelp

Description: A huge and impressive kelp that grows in deep water. The stipe, 7–27 m long, terminates in a large pneumatocyst 15–20 cm in diameter. Two flattened branches arise in one plane off the top of the float; the branches divide dichotomously and each terminates in a large single blade 6–20 m long and up to 1 m wide.

Pacific Coast Distribution: Point Conception, California, to Baja California.

Habitat: On rocks and gravel, subtidal, 30 to 90 m.

Remarks: Even though *Pelagophycus* is a very large kelp it grows in deep enough water that the blades rarely reach the surface. It is not uncommon, however, to find the large pneumatocyst, bearing two branches or “antlers,” floating offshore or washed up on the beach.

Order Fucales

Family Fucaceae

Fucus gardneri Silva — Rockweed, Popweed, Bladderwrack

Description: A discoid holdfast supports a flattened, dichotomously branched thallus with a distinct midrib. The base of the thallus becomes stipe-like through abrasion of the blade that erodes to the midrib. At maturity, the blades develop swollen tips called receptacles that contain air and mucilage. A cross section of a receptacle reveals multiple chambers called conceptacles where gametes are formed.

Pacific Coast Distribution: Bering Sea and Aleutian Islands, AK, to Santa Barbara County, CA.

Previous names: *Fucus distichus* Linnaeus

Habitat: Abundant, on rocks and mussels, in exposed to sheltered habitats, mid to low intertidal.

Remarks: In the northeastern Pacific, *Fucus* is often the dominant intertidal algae and forms a dense band in the mid to low intertidal.

D. Richards

Hesperophycus californicus thallus.

D. Richards

Close-up of *H. californicus* branches.

Hesperophycus californicus Silva

Description: A olive-green, dichotomously branched thallus, 10–50 cm tall. The branches have a distinct midrib that is lined on either side by two parallel rows of tiny hairs, called cryptostomata. Mature plants will have swollen receptacles at the tips of the branches.

Pacific Coast Distribution: Santa Cruz, California, to northern Pacific Mexico.

Previous names: *Hesperophycus harveyanus* (Decaisne) Setchell *et* Gardner

Habitat: Locally abundant, on rocky headlands, upper to mid intertidal.

Remarks: This species can be found mixed with *Fucus* in central California. In southern California, however, *Hesperophycus* replaces *Fucus*.

Pelvetiopsis limitata Gardner Dwarf Rockweed

Description: The flattened, dichotomously branched thallus has no midrib, is olive green to light tan in color, and 8–15 cm tall. Inflated, warty receptacles (5–15 mm long) are present on the tips of mature branches.

Pacific Coast Distribution: Hope Island, British Columbia, to San Luis Obispo County, California.

Habitat: Common, on rocks, in the high intertidal, in exposed habitats.

Remarks: Named after the French botanist, Dr. Pelvet.

Silvetia compressa (Agardh) Serrao, Cho, Boo, *et* Brawley

Description: A leathery conical holdfast produces one to several dichotomously branched axes. The narrow branches are oval to flattened in cross section, have no midrib and, when mature, bear warty receptacles on the tips. The olive-green thallus can get up to 60 cm long.

Pacific Coast Distribution: Shelter Cove, California, to Baja California.

Previous names: *Pelvetia compressa* (Agardh) DeToni, *Pelvetia fastigiata* (J. Agardh) DeToni, *Fucus compressus* C. Agardh

Habitat: Common, on rocks in exposed habitats, in the high intertidal.

Family Cystoseiraceae

Cystoseria geminata thallus.

Close-up of floats on *Cystoseria geminata.*

Cystoseira geminata C. Agardh Northern Bladder Chain Kelp

Description: The thallus consists of a discoid holdfast, stipe, and one to several main branches. The main branches produce numerous alternating side branches that ultimately bear many short terminal branchlets and spherical pneumatocysts that have a pointed projections at their apex. The entire plant is dark brown in color and can reach a height of 2–5 m.

Pacific Coast Distribution: Bering Sea and Aleutian Islands, AK, to Whidbey Island, Washington.

Habitat: Common, on rocks, in low intertidal and upper subtidal, in sheltered habitats.

Remarks: Can easily be confused with *Sargassum muticum*, however the pointed tips on the pneumatocysts distinguish *Cystoseira*.

Cystoseria osmundacea thallus.

Close-up of floats on *Cystoseria osmundacea.*

Cystoseira osmundacea (Turner) C. Agardh Bladder Chain Kelp

Description: A disc-shaped holdfast gives rise to a woody, erect stipe that is triangular in cross-section. Flattened, pinnate branches with a slight midrib arise radially off the stipe. The upper portions of the branches become cylindrical and bear small (7 mm diameter), spherical pneumatocysts that can be single, in pairs, or in chains and have pointed projections at their apex. The thallus can get up to 8 m in length.

Pacific Coast Distribution: Seaside, Oregon, to Baja California.

Previous names: Cystoseira expansa C. Agardh

Habitat: Common, in pools and on rocks, in the low intertidal and shallow subtidal.

Remarks: The lower position of the thallus persists throughout the year and the upper portion of the thallus breaks off.

Family Sargassaceae

Sargassum muticum (Yendo) Fensholt Wireweed

Description: The main axes, anchored by a discoid holdfast, branches repeatedly and alternately to form a wiry thallus that is golden brown in color and up to 2 m tall. Each branch bears small (up to 4 cm) elongated blades and spherical floats that can occur singly or in small clusters.

Pacific Coast Distribution: Prince of Wales Island, Alaska, to Baja California; native to Japan and China.

Habitat: Common to locally abundant, on rocks and mud, in the low intertidal and upper subtidal; found in sheltered to moderately exposed habitats.

Remarks: Introduced to Puget Sound, Washington, from Japan, probably on Japanese oysters in the 1930's, *S. muticum* has since spread up and down the west coast. This species was also introduced to France and Britain where it is reportedly causing displacement of native algal species through over-growing and shading.

Some tropical species of *Sargassum* are always free floating. Portuguese sailors encountered huge rafts of this algae in the Atlantic, west of Africa, and named the area the Sargasso Sea after a type of grape, *sarga*.

PHYLUM RHODOPHYTA
Class Rhodophyceae

ORDER COMPSOPOGONALES
Family Erythrotrichiaceae

Smithora naiadum (Anderson) Hollenberg
Description: The thallus consists of thin purplish-red blades that grow epiphytically on seagrasses. The blades reach about 5 cm in length and grow from cushion-like holdfast.
Pacific Coast Distribution: Kodiak Island, Alaska, to Isla Magdalena, Baja California.
Previous names: Porphyra naiadum Anderson, *Smithora naiadum* var. *australis* Dawson
Habitat: On blades of *Phyllospadix* and *Zostera*, low intertidal and upper subtidal.
Remarks: *Porphyra* can also be found growing on blades of seagrasses; accurately differentiating *Porphyra* and *Smithora* requires microscopic examination.

ORDER BANGIALES
Family Bangiaceae

***Porphyra* spp.**
Nori, Black Seaweed, Laver
Description: The thallus consists of a thin, delicate blade that is only one or two cell layers thick in cross section (you can read a book through a blade!). Species of *Porphyra* range in color from dark purple to yellowish green to reddish pink and the blades will be almost black when they are dry.
Pacific Coast Distribution: Bering Sea and Aleutian Islands, Alaska, to Baja California.
Habitat: On rocks, man-made structures and other algae; throughout the intertidal and upper subtidal.
Remarks: At least 20 species of *Porphyra* are found along our coast. Distinguishing species can be quite difficult; they are differentiated by characteristics such as the shape of the blade, thallus color and microscopic features like the reproductive cells. *Porphyra* is one of the most extensively cultivated seaweeds in the world. This tasty edible can be pressed into sheets and used to wrap sushi or eaten in a host of other dishes.

Order Ahnfeltiales
Family Ahnfeltiaceae

Ahnfeltia fastigiata (Postels *et* Ruprecht) Makienko
Description: A stiff and wiry alga up to 20 cm tall, composed of dense axes that arise from a matted rhizome-like holdfast. Axes are cylindrical and repeatedly dichotomously branched. The thallus is dark purple to black in color.
Pacific Coast Distribution: Aleutian Islands, Alaska, to northern Baja California.
Previous names: This species has been improperly identified along the Pacific coast as *Ahnfeltia plicata* (which is found in the Atlantic).
Habitat: Common, on rocks, often in sandy places, mid to low intertidal.
Remarks: *Ahnfeltia* can survive periods of sand burial. A similar species, *A. plicata*, is harvested in the north Atlantic and along Russian coasts for agar production.

Order Corallinales
Family Corallinaceae

***Calliarthron* spp.**
Description: An articulated coralline alga, up to 25 cm tall, with flattened intergenicula that bear conceptacles along their edges. Branching can be dichotomous or irregular or pinnate.
Pacific Coast Distribution: Northern southeast Alaska to Baja California.
Habitat: On rocks, low intertidal and subtidal.
Remarks: Several species of *Bossiella* have similar morphology to *Calliarthron*. The location of the conceptacles along the margins of the intergenicula is a distinguishing characteristic for members of *Calliarthron*.

Pseudolithophyllum spp.
Description: A purplish-pink crustose coralline thallus that may be smooth or covered with white, rounded protuberances.
Pacific Coast Distribution: Alaska to Mexico
Habitat: Abundant, on rocks and mollusc shells (especially the Whitecap Limpet, *Acmaea mitra*), low intertidal and subtidal.
Remarks: There are at least five genera and nine species of encrusting coralline algae along our coast and they cover more area in the low intertidal and shallow subtidal than any other group of organisms. Species can be difficult to correctly identify in the field; they are distinguished by characteristics such as thickness of crust, cell dimensions, and reproductive structures.

Melobesia mediocris (Foslie) Setchell *et* Mason
Description: A pinkish, coralline crust that grows in circular patches on seagrasses. Patches grow to about 5 mm in diameter and can become irregular in shape when they are densely clustered.
Pacific Coast Distribution: Queen Charlotte Islands, British Columbia, to Baja California.
Previous names: Lithophyllum zostericola f. *mediocre* Foslie; *Melobesia zostericola* f. *mediocre* (Foslie) Foslie; *Lithothamnion mediocre* (Foslie) Foslie *et* Nichols.
Habitat: On blades of *Phyllospadix* and *Zostera*, low intertidal and upper subtidal.
Remarks: A closely related species, *Melobesia marginata*, has the same distribution along the coast but grows epiphytically on red algae including: *Osmundea*, *Ahnfeltia* and *Rhodymenia*.

Serraticardia macmillanii (Yendo) Silva
Description: An articulated coralline alga up to 12 cm tall with regular pinnate branching. The upper axial intergenicula are wider than they are tall, have a hexagonal outline, and usually bear branches. The lower portion of the thallus is often devoid of branches.
Pacific Coast Distribution: Cook Inlet, Alaska, to Channel Islands, California.
Previous names: Cheilosporum macmillanii Yendo; *Calliarthron pinnulatum* Manza
Habitat: On rocks, in very low intertidal and subtidal.

Bossiella spp.

Description: Members of this genus have an upright thallus consisting of a series of calcified intergenicula separated by short non-calcified genicula. Geniculum is Latin for knee. The presence of the non-calcified "knees" characterizes the genera known as the articulated corallines from the crustose corallines. *Bossiella* can be up to 20 cm tall and the thallus is flattened and dichotomously (occasionally pinnately) branched.
Pacific Coast Distribution: Bering Sea, Alaska, to Baja California, Mexico.
Habitat: On rocks, low intertidal and subtidal.
Remarks: Although similar in appearance to *Calliarthron*, *Bossiella* is easily distinguished by the location of the conceptacles on the flat surfaces of the intergenicula in contrast to the conceptacles on the margins of the intergenicula in *Calliarthron*.

Corallina vancouveriensis Yendo

Description: A profusely branched, articulated coralline algae with pinnately arranged, erect branches. The calcified segments are cylindrical and the conceptacles are found on the tips of the branches.
Pacific Coast Distribution: Aleutian Islands, Alaska, to Baja California; Galápagos Islands.
Previous names: *Corallina aculeata* Yendo; *C. gracilis* f. *densa* Collins; *C. densa* (Collins) Doty
Habitat: On rocks, mid to low intertidal and subtidal.
Remarks: This species can form dense mats in the low intertidal in exposed habitats.

Family Galaxauraceae

Scinaia confusa (Setchell) Huisman

Description: The thallus is bright to dark red, dichotomously branched, and 5 to 20 cm tall. Branches are cylindrical and have tapering tips.
Habitat: Uncommon to fairly common, on rocks, low intertidal and subtidal.
Pacific Coast Distribution: Kruzof Island, Alaska, to Punta San Quintin, Baja California (also known from Costa Rica and Isla Isabela, Galápagos).
Previous name: Pseudogloiophloea confusa (Setchell) Levring
Remarks: This species exhibits alternation of heteromorphic generations. The gametophyte is a macroscopic annual. The tetrasporophyte is microscopic and, although it has been grown in culture, it is unknown from nature.

Cumagloia andersonii (Farlow) Setchell *et* Gardner

Description: A soft, almost gelatinous, alga that is elastic but quite tough. Thallus consists of one or more axes that arise from a small disc-shaped holdfast. Axes are cylindrical to oval in cross section and covered with small, spinelike branches. The thallus is dark brown to purplish red in color and 10 to 30 cm tall.
Pacific Coast Distribution: Cold Bay, Alaska Peninsula, to Baja California.
Habitat: On rocks, usually in exposed habitats, upper to mid intertidal.
Remarks: Extracts from this species demonstrate anti-viral activity.

Family Palmariaceae

Halosaccion glandiforme (Gmelin) Ruprecht Sea Sacs

Description: The thalli consist of one or more erect, hollow sacs that arise from a small discoid holdfast. Sacs are usually 5 to 15 cm tall and 3 to 4 cm in diameter. Color ranges from yellowish-brown, if the thallus is living in full sunlight, to reddish-purple if found growing in shade. The hollow portion of the thallus is usually filled with seawater and a small air bubble at the tip. As the plants get older, however, the tips of the sacs can erode and leave the thalli flat or filled with sand.
Habitat: Abundant, on rocks and other algae, upper to low intertidal.
Pacific Coast Distribution: Aleutian Islands, Alaska, to Point Conception, California.
Remarks: In 1982, Lee found differences in cellular and reproductive structures in *Halosaccion* and named a new species, *H. americanum* Lee. A more recent study (Lindstrom *et* al. 1996), supports previous work that determined it is not possible to distinguish between these two species. For now, the taxonomy of *H. americanum* is uncertain.

Palmaria callophylloides Hawkes *et* Scagel

Description: The thallus consists of repeatedly dichotomously branching blades. The thallus is usually about 15–25 cm tall, has a crisp texture and is deep red to reddish brown in color.
Pacific Coast Distribution: Shemya Island (Aleutian Islands), AK, to northern British Columbia.
Previous names: Palmaria palmata f. *sarniensis*
Habitat: On rocks, mid to low intertidal.
Remarks: The specific epithet, *callophylloides*, refers to this alga's resemblance to *Callophyllis*.

Palmaria hecatensis Hawkes

Description: The thallus, up to 40 cm tall, consists of a broad blade that is dichotomously divided once into 2 rounded lobes. The blades are deep red to maroon in color and have a thick, strap-like texture.

Pacific Coast Distribution: Shemya Island (Aleutian Islands), Alaska, to Humboldt County, California.

Previous names: This species was previously misidentified as *Rhodymenia palmata* f. *palmata*, a North Atlantic species commonly called Dulse that is now know as *Palmaria palmata.*

Habitat: On rocks, mid to low intertidal.

Remarks: This species is a versatile edible; a very similar species, *Palmaria palmata,* is probably the most commonly eaten North Atlantic seaweed.

Palmaria mollis (Setchell *et* Gardner) van der Meer *et* Bird

Description: A wedge-shaped blade (narrow at the base) that is irregularly split into lobes with jagged tips. The blades are dull red in color, somewhat flaccid in texture and can bear marginal bladelets.

Pacific Coast Distribution: Attu Island (Aleutian Islands), AK, to Santa Barbara County, California.

Previous names: Rhodymenia palmata f. *mollis* Setchell *et* Gardner; *Palmaria palmata* f. *mollis* (Setchell *et* Gardner) Guiry

Habitat: On rocks and kelps, low intertidal and upper subtidal.

Remarks: This species grows in clumps and the second year blades grow from the same base.

Order Ceramiales
Family Ceramiaceae

Callithamnion pikeanum Harvey

Description: A profusely and radially branched thallus with one or more obvious axes that arise from a common discoid holdfast. The thallus, 10–20 cm tall, has a wooly appearance when it is dry and is reddish brown in color; the dense branches are often covered with diatoms.
Pacific Coast Distribution: Attu Island (Aleutian Islands), Alaska, to Los Angeles County, California.
Previous names: Ceratothamnion pikeanum f. *laxum* Setchell *et* Gardner, *Callithamnion laxum* (Setchell *et* Gardner) Setchell *et* Gardner.
Habitat: Common, on rocks, throughout the intertidal region.
Remarks: Most of the members of this genus are very small and filamentous and can only be identified via microscopic examination, *Callithamnion pikeanum* is the only species that develops thickened axes and branches.

Ptilota and *Neoptilota*

Description: The members of these two genera have flattened, feathery thalli with regularly opposite branching off straight main axes. One branch from each opposing pair is very small in comparison to the other. The ultimate branchlets can have smooth edges or be serrated on one or both sides.
Pacific Coast Distribution: Bering Sea and Aleutian Islands, Alaska, to Baja California.
Habitat: On rocks and other algae, low intertidal and subtidal, usually in exposed habitats.
Remarks: There are at least six species in these two genera of beautiful, delicately branched algae; distinguishing these species requires excellent eyes, a hand lens or a dissecting microscope. They are differentiated by the presence or absence of serrations on the small ultimate branchlets, the amount of growth in oppositely paired branches, and the location of reproductive structures.

Ceramium spp.

Description: These delicate, dichotomously branched filaments consist of colorless central cells that are totally or partially covered by a layer of tiny, red cells. In several species the red corticating cells do not cover the entire central cell, giving a distinctive banded appearance to the thallus when observed under a hand lens. The tips of the branches are forked and look like tiny crab pincers.
Pacific Coast Distribution: Alaska to Baja California.
Habitat: On rocks and man-made structures and other algae, throughout the intertidal region.
Remarks: There are many species of *Ceramium* along our coast; one of them, *C. codicola*, lives exclusively on *Codium fragile*.

Microcladia borealis Ruprecht

Description: A delicate, profusely branched thallus with successive orders of unilateral branching (branches grow only on one side of the axis, like teeth on a comb). The branches are curved and have forked tips. The thallus is deep to rose red in color and attaches to the substrate via rhizoidal branches.
Pacific Coast Distribution: Aleutian Islands, Alaska, to San Luis Obispo County, California.
Habitat: Abundant, on rocks and epiphytic on kelp, mid to low intertidal and upper subtidal.

Microcladia coulteri Harvey

Description: A flattened, rose-red alga with regularly alternate branching that occurs on each side of a straight axis. The thallus can get up to 25 cm, but is usually much smaller.
Pacific Coast Distribution: Vancouver Island, British Columbia, to Punta San Rosalia, Baja California.
Habitat: Obligate epiphyte on various red algae including *Chondracanthus* and *Mazzaella*. Shown here on a piece of *Prionitis*.
Remarks: Although most epiphytic algae simply grow on top of their hosts, *M. coulteri* produce small filaments that penetrate their host tissue. Another species, *M. californica*, is an obligate epiphyte on *Egregia* and only occurs on the surface of the kelp. Studies of these host-epiphyte interactions have shed light on the evolution of red-algal parasitism.

Family Delesseriaceae

Delesseria decipiens J. Agardh
Description: The thallus consists of delicate pinkish- to purplish-red blades, about 15 mm wide, with a prominent midrib and alternate branching. A distinguishing feature is the presence of small leaflets that arise directly from the midrib.
Pacific Coast Distribution: Aleutian Islands and Prince William Sound, Alaska, to San Luis Obispo, California.
Habitat: Common, on rocks, especially on the underside of overhanging rocks, low intertidal and subtidal.
Remarks: This species has isomorphic alternation of generations. The tetrasporangia are found along the midrib; male and female gametangia are on separate thalli, cystocarps are borne on fertile leaflets, the spermatangia are scattered over the blade.

Cryptopleura ruprechtiana (J. Agardh) Kylin
Description: A deep- to brownish-red, flattened alga that is divided into many fan-shaped blades. The edges of the blades are usually covered with small, frilly ruffles. A midrib can usually be seen a the base of the blade and this gives way to a network of very fine veins (you may have to hold the blade up to the light to see them).
Pacific Coast Distribution: Northern Southeast Alaska to Punta María, Baja California.
Previous names: Nitophyllum ruprechtianum J. Agardh; *Cryptopleura farlowianum* var. *farlowianum* Ver Steeg *et* Josselyn; *Botryoglossum farlowianum* (J. Agardh) DeToni; *Botryoglossum ruprechtianum* (J. Agardh) DeToni
Habitat: Common, on rocks, low intertidal and subtidal.
Remarks: There is a question as to whether this species should be placed in the genus *Botryoglossum* or *Cryptopleura* and if there is a separate species, *C. farlowiana*, in California.

Hymenena flabelligera (J. Agardh) Kylin
Description: A rose- to brownish-red thallus divided into many fan-shaped blades with a network of very fine veins. Tetrasporangia located on the blade in linear series.
Pacific Coast Distribution: Southeast Alaska to San Luis Obispo, California.
Previous names: *Hymenena kylinii* Gardner
Habitat: Common, on rocks, in the low intertidal and subtidal.
Remarks: It can be very challenging to distinguish *Hymenena* and *Cryptopleura*. The genera are differentiated by the location of the tetrasporangia on the blade, but if you don't have a tetrasporophyte this is obviously difficult to determine!

Membranoptera platyphylla (Setchell *et* Gardner) Kylin
Description: The thallus, up to 8 cm tall, consists of rose-red to deep-red, thin blades that arise from a discoid holdfast. The blades have an irregular dichotomous branching arrangement, a distinct midrib and delicate, pinnate veins.
Pacific Coast Distribution: Prince William Sound, Alaska, to Channel Islands, California.
Previous names: Pteridium serratum f. *platyphyllum* Setchell *et* Gardner; *Membranoptera multiramosa* Gardner
Habitat: Common, on pilings, rocks and stipes of kelp, in the low intertidal and upper subtidal.

Polyneura latissima (Harvey) Kylin
Description: A purplish- to rose-red thallus (12–25 cm tall) composed of one or more blades that may be irregularly divided. The blades are thin, have a crinkly texture, and are covered with a conspicuous network of anastomosing veins.
Pacific Coast Distribution: Prince William Sound, Alaska, to Baja California.
Previous names: Hymenena latissima Harvey
Habitat: Frequent, on rocks, mid to low intertidal and subtidal.
Remarks: The fact that *Polyneura* means "many nerves or veins" makes it easy to remember the descriptive name of this alga. Extracts from *Polyneura* have been shown to inhibit viruses.

Family Rhodomelaceae

Neorhodomela larix (Turner) Masuda
Description: The thallus, dark-brown to black in color, is composed of a clump of cylindrical axes that have a dense covering of short, radially arranged branchlets. The axes, which can be up to 30 cm long, are often twisted into spirals.
Pacific Coast Distribution: Bering Sea and Aleutian Islands, Alaska, to Baja California, Mexico.
Previous names: Rhodomela larix Turner
Habitat: Abundant, on rocks, throughout the intertidal.
Remarks: This algae can form dense mats on rocks and is often associated with *Odonthalia*.

Odonthalia floccosa (Esper) Falkenberg
Description: A dark-brown, profusely branched alga, up to 40 cm tall, that attaches to the substrate by a small disk-shaped holdfast. Branches are cylindrical, alternately arranged and bear clusters of short, pointed branchlets.
Pacific Coast Distribution: Aleutian Islands and Bering Sea, Alaska, to Santa Barbara County, California.
Habitat: Abundant, on rocks, throughout the intertidal.
Remarks: There are at least four other species of *Odonthalia* along our coast, some of which have flattened branches; the presence of the clusters of pointed branchlets is usually characteristic of all these algae. The saccate brown alga, *Soranthera*, is a common epiphyte on *Odonthalia.*

Osmundea spectabilis (Postels *et* Ruprecht) Nam

Description: The thallus consists of a dense cluster of axes that bear somewhat flattened, blunt-tipped, opposite branches. The alga is purplish-red in color and has a firm, tough and elastic texture.

Pacific Coast Distribution: Sitka, Alaska, to Isla Guadalupe, Baja California, Mexico.

Previous names: Laurencia spectabilis Postels *et* Ruprecht

Habitat: Common, on rocks, in the mid to low intertidal.

Remarks: This alga is sometimes called Pepper Dulse due to its spicy flavor. *Osmundea* has a distinct, pungent smell that results from the presence of bromine- and chlorine- containing compounds. In a closely related species, *Laurencia obtusa*, these compounds show significant antibacterial activity.

Polysiphonia and ***Pterosiphonia***

Description: The members of these two genera have very fine, delicately branched thalli, that are bright red to dark brown in color and form clumps or mats. The structural details of the branches needs to be viewed with a microscope; branches consist of a filament of central axial cells each one of which is surrounded by equal length cells called pericentral cells. When viewed from the side, these pericentral cells look like tiers of bricks.

Pacific Coast Distribution: Aleutian Islands, Alaska, to Baja California.

Habitat: On rocks, other algae, and man-made structures, in the mid to low intertidal and subtidal.

Remarks: Along the Pacific coast, there are at least 14 species (and lots of varieties!) that belong to these two genera. Examination with a microscope is necessary for accurate identification.

Order Gelidiales
Family Gelidiaceae

Gelidium and *Pterocladia*

Description: These two genera have highly variable, branched thalli that are rigid and tough in texture, a red to deep purple or black in color, and up to 50 cm tall. The branching arrangement is often in two rows off a central axis, but it can also be irregular. The axes are slender (about 3 mm wide) and are usually cylindrical or oval (*Gelidium*), or flattened (*Pterocladia*), when viewed in cross section.
Pacific Coast Distribution: Southern southeast Alaska to Baja California.
Habitat: On rocks, low intertidal and shallow subtidal.
Remarks: There are about 10 representatives of these two genera along our coast, variable morphology and resemblance to each other can make species identification difficult. Members of both of these genera are excellent sources of agar.

Order Gigartinales
Family Dumontiaceae

Constantinea simplex Setchell

Cup and Saucer Seaweed
Description: This alga is easily recognized as a cup-shaped blade supported by a unbranched, cylindrical stipe attached in the center of the blade. The thallus is perennial and each year the stipe projects through the center of the blade and a new blade grows directly above the old blade; eventually the old blade wears away and leaves a scar. The blade can be up to 10 cm in diameter, is dark red in color, and has a leathery texture.
Pacific Coast Distribution: Kodiak Island and Aleutian Islands, Alaska, to Point Conception, California.
Habitat: On rocks, low intertidal and subtidal.
Remarks: By counting the annual blade scars on the stipe of *Constantinea* you can calculate the age of the thallus. Extracts from this seaweed have been shown to inhibit several viruses.

Constantinea rosa-marina (Gmelin) Postels *et* Ruprecht Northern Cup and Saucer Seaweed
Description: A deep-red thallus with a cylindrical, branched stipe. Each branch terminates in the center of a cup-shaped blade. The blades can be up to 20 cm in diameter.
Previous names: Constantinea sitchensis Postels *et* Ruprecht
Pacific Coast Distribution: Bering Sea to southeastern Alaska.
Habitat: On rocks, extreme low intertidal and subtidal.
Remarks: Another species, *Constantinea subulifera*, also has a branched stipe and ranges from the Aleutian Islands, Alaska, to Washington. The cup-shaped blade of *C. subulifera* can be up to 35 cm in diameter, has conspicuous veins, and the stipe usually protrudes through the center of the blade.

Cryptosiphonia woodii (J. Agardh) J. Agardh
Description: A profusely branched, small (10 to 25 cm tall), species that occurs in dense tufts. The branches are cylindrical and have an irregular, radial arrangement. The thallus is usually dark brown to deep maroon in color, however, late in the summer, it can be yellowish to olive-brown.
Pacific Coast Distribution: Unalaska Island, Alaska, to San Pedro, California.
Previous names: Pikea woodii (J. Agardh)*; Cryptosiphonia grayana* (J. Agardh) J. Agardh
Habitat: Abundant, on rocks, in the mid intertidal region.
Remarks: Researchers have found polysaccharide extracts from *Cryptosiphonia woodii* that are particularly useful as anti-herpes substances.

Pikea californica Harvey

Description: The thallus, 5-10 cm tall and deep red in color, has several fronds arising from a small disk-shaped holdfast. The axes are flattened, 1–2 mm wide and have an opposite branching arrangement. The tips of the branches often are pointed and spinelike.

Pacific Coast Distribution: Prince William Sound, AK, to Isla San Martín, Baja California.

Habitat: Common, on rocks, in the low intertidal and subtidal.

Remarks: This species is native to our coast but it is also found on the Isle of Scilly, England, where it is non-native. It is not know how *Pikea* arrived in England, but one theory is that the alga was inadvertently brought from California on military equipment during World War II.

Dilsea californica (J. Agardh) Kuntze

Description: The thallus is a maroon to brownish red blade (about 15 cm tall) that is often deeply split into many curved or sickle-shaped pieces. The blade has a thick, almost leathery texture, it is not elastic, and often has a slightly muted, dull finish.

Pacific Coast Distribution: Attu Island (Aleutian Islands), Alaska, to San Luis Obispo County, California.

Previous names: Sarcophyllis californica J. Agardh

Habitat: On rocks, in the low intertidal and subtidal.

Remarks: *Dilsea* and closely related *Neodilsea* are difficult to accurately identify in the field. Examination of a blade cross section reveals the characteristics needed to confirm the identity of these bladed red algae.

Farlowia compressa (J. Agardh)

Description: This species is coarse in texture, dark-red in color and up to 50 cm tall. The thallus consists of 3–12 mm wide, flattened branches. A thickened midline that somewhat resembles a midrib can be present on the main branches.

Pacific Coast Distribution: Fort Bragg, California, to Punto Santo Tomás, Baja California.

Habitat: On rocks, low intertidal and subtidal.

Remarks: This is considered a wide form of *Farlowia mollis* by some phycologists.

Farlowia mollis (Harvey *et* Bailey) Farlow *et* Setchell
Description: A rather tattered looking alga that is up to 20 cm tall, red to blackish-red in color and consists of flattened axes 1–4 mm wide that have an alternate branching pattern. The higher order branches are almost cylindrical and quite thin.
Pacific Coast Distribution: Prince William Sound, Alaska, to Isla San Martín, Baja California.
Habitat: Common, on rocks, often in sandy habitats, mid to low intertidal and subtidal.
Remarks: *Farlowia* has heteromorphic alternation of generations. A crustose tetrasporophyte alternates with the upright gametophyte that is shown here.

Family Endocladiaceae

Endocladia muricata (Postels *et* Ruprecht) J. Agardh
Description: A profusely branched, wiry thallus that forms dark-brown to reddish-brown clumps that are 3 to 8 cm tall. The branches are cylindrical and covered with short spines.
Pacific Coast Distribution: Amchitka Island (Aleutian Islands), Alaska, to Punta Santo Tomás, Baja California.
Habitat: Abundant, on rocks, in the high intertidal.
Remarks: The highly branched, bushy nature of this alga enables it to retain moisture during low tide. *Endocladia* is usually found in association with the small barnacle, *Balanus*; it is also a preferred settling site for mussels.

Gloiopeltis furcata covering boulders in the high intertidal.

Close-up of *G. furcata* thallus.

Gloiopeltis furcata (Postels *et* Ruprecht) J. Agardh
Description: A irregularly dichotomously branched thallus that grows from an encrusting base and forms tufts 2 to 5 cm tall. The cylindrical branches are smooth. The thallus is reddish-brown in color and has a gelatinous texture when wet; it becomes "crunchy" and almost black when dry.
Pacific Coast Distribution: Aleutian Islands, Alaska, to Punta Eugenia, Baja California.
Previous names: *Dumontia furcata* Postels *et* Ruprecht
Habitat: Abundant, on rocks, in the high intertidal.
Remarks: The upright branches of *Gloiopeltis* are annual and they grow from the perennial base each spring.

Family Furcellariaceae

Opuntiella californica (Farlow) Kylin
Description: The thallus is composed of roughly circular, undivided blades that produce secondary blades from their margins. The blades, which are quite thick and have a tough texture, are dark red in color.
Pacific Coast Distribution: Attu Island, Aleutian Islands, Alaska, to Baja California.
Habitat: On rocks, extreme low intertidal and subtidal, in exposed habitats.
Remarks: This alga was named after the prickly pear cactus, *Opuntia,* due to the superficial similarity in their branching patterns. *Opuntiella* has heteromorphic alternation of generations; the gametophyte alternates with a crustose tetrasporophyte that was previously described as *Cruoria profunda*.

Family Gigartinaceae

Chondracanthus exasperatus (Harvey *et* Bailey) Hughey Turkish Towel
Description: A purplish-red, tough blade that is covered with numerous bumps. The blade, up to 50 cm tall and 18 cm wide, is usually lanceolate (tapering toward each end) but, especially in the northern part of its range, it can be somewhat irregular and jagged in outline.
Pacific Coast Distribution: Yakutat Bay, Alaska, to Baja California.
Previous names: *Gigartina exasperata* Harvey *et* Bailey; *G. californica* J. Agardh
Habitat: Common, on rocks, low intertidal and subtidal.
Remarks: Some phycologists believe that the regularly shaped, lanceolate bladed form is a separate species, *C. californicus*.

Chondracanthus canaliculatus on intertidal wash rock.

Close-up of *C. canaliculatus* showing branching patterm.

Chondracanthus canaliculatus (Harvey) Guiry
Description: The thallus is somewhat flattened, 10–15 cm tall, and has an irregular opposite branching pattern. Branches are reddish-brown to yellowish-brown at the tips.
Pacific Coast Distribution: Yaquina Head, Oregon, to Isla Magdalena, Baja California.
Previous names: *Gigartina serrata* Gardner; *G. canaliculatus* Harvey
Habitat: Common to abundant, on rocks, in the mid to low intertidal.
Remarks: All members of the genus *Chondracanthus* have isomorphic alternating generations.

Chondracanthus corymbiferus (Kützing) Guiry

Description: A very large (up to 1 m), tough blade that is oval in outline and covered with course bumps. When thalli are in the low intertidal, the color of the blades is yellowish-pink; subtidal thalli are bluish-red.

Pacific Coast Distribution: Barkley Sound (Vancouver Island), British Columbia, to Cabo San Quintin, Baja California.

Previous names: Gigartina corymbifera (Kützing) J. Agardh; *G. binghamiae* J. Agardh

Habitat: Common, on rocks in exposed locations, low intertidal and subtidal.

Remarks: Various species of *Chondracanthus* are harvested in Chile for carrageenan production.

Mazzaella affinis (Harvey) Fredericq

Description: A repeatedly and dichotomously branched alga that grows to be 4–10 cm tall. The thalli are reddish-brown to purple-black in color and grow in dense clumps.

Pacific Coast Distribution: Bering Strait, Alaska, to Baja California.

Previous names: Rhodoglossum affine (Harvey) Kylin; *Gigartina affine* (Harvey) Kim

Habitat: Locally common, on rocks in mid to low intertidal.

Remarks: *Mazzaella* is closely related to the Atlantic alga, *Chondrus crispus*, known as Irish Moss, or Carrageen, that is used to make soups and jellies.

Mazzaella oregona (Doty) Hughey, Silva *et* Hommersand

Description: The thallus is composed of a clump of cleft or irregularly shaped blades (10–15 cm tall). The blades usually have wavy margins, are flaccid and stretchy in texture, and reddish-brown in color. Cystocarps are large (up to 4 mm in diameter) and bulge from the surface of the blades.

Pacific Coast Distribution: Southern Alaska to Ventura, California.

Previous names: Iridophycus oregonum Doty; *Iridaea heterocarpa* Postels *et* Ruprecht; *Gigartina heterocarpa* Kim; *Mazzaella heterocarpa* Fredericq

Habitat: Common, on rocks in exposed habitats, mid to upper intertidal in southern portion of range, low intertidal in British Columbia and southern Alaska.

Remarks: A very similar species, *Mazzaella phyllocarpa,* is found in Alaska.

Mazzaella parksii (Setchell *et* Gardner) Hughey, Silva *et* Hommersand
Description: The thallus is a dense cluster of small (2–4 cm tall) blades, yellowish-green to brownish-purple in color, that arise from a perennial encrusting holdfast. The blades may be divided into two or three rounded lobes or they may undivided and shaped like tiny Chinese soup spoons.
Pacific Coast Distribution: Amchitka Island, Aleutian Islands, Alaska, to Bodega Head, Sonoma County, California.
Previous names: Iridophycus parksii Setchell *et* Gardner*; Iridaea cornucopiae* Postels *et* Ruprecht; *Gigartina cornucopiae* (P. *et* R.) Kim; *Mazzaella cornucopiae* (P. *et* R.) Hommersand
Habitat: Locally abundant on exposed headlands, in the high intertidal.
Remarks: Cystocarps are large and bulge from the surface of the blades.

Mazzaella splendens (Setchell *et* Gardner) Fredericq
Description: The thallus is composed of a cluster of blades that are usually iridescent dark-purple and brown, but may appear black. Blades arise from short (2–4 cm long) stipes that have small, discoid holdfasts. The shape of the blades varies with habitat and the amount of wave exposure, blades can be narrowly or broadly lanceolate, heart-shaped, or oval.
Pacific Coast Distribution: Southeast Alaska to northern Baja California.
Previous names: Iridaea splendens (Setchell *et* Gardner) Papenfuss; *Iridaea cordata* var. *splendens* (S. *et* G.) Abbott
Habitat: Abundant, on rocks, mid to low intertidal and upper subtidal. Found in both exposed and moderately sheltered habitats.
Remarks: Another species, *Mazzaella flaccida*, is very similar to *M. splendens* and be difficult to distinguish in the field. Generally, *M. flaccida* is yellowish-green and only iridescent-purple near the base of the blade.

Callophyllis flabellulata Harvey
Description: A flattened, rose- to orange-red, thin bladed alga (up to 10 cm tall) that is roughly fan-shape in outline. The thallus is divided into many ribbonlike blades that have an irregular to pinnate branching arrangement and irregularly notched tips.
Pacific Coast Distribution: Amchitka Island, Aleutian Islands, Alaska, to Punta Velero, Baja California.
Previous names: Callophyllis marginifructa Setchell *et* Swezy; *C. acrocarpa* Setchell; *C. crassifolia* Setchell *et* Swezy; *C. filicina* Setchell *et* Swezy; *C. gardneri* Setchell; *C. odonthalioides* Setchell
Habitat: Common, on rocks and other algae, low intertidal and subtidal.
Remarks: Extracts from this alga have been shown to have active antiviral properties.

Callophyllis pinnata Setchell *et* Swezy
Description: The dark red thallus is shaped like a hand, with 10–30 cm long dichotomously branched blades. The branches are about 2–3 cm wide and have somewhat pointed tips.
Pacific Coast Distribution: Kayak Island, Alaska, to Baja California.
Habitat: Locally abundant, on rocks and epiphytic on kelp stipes, low intertidal and subtidal.
Remarks: The name *Callophyllis* is based on the Greek roots, *kallo* and *phyllo*, which mean beautiful and leaf respectively.

New *Erythrophyllum delesserioides* blades in early spring.

E. delesserioides blades arising from a remnant midrib.

Erythrophyllum delesserioides J. Agardh
Description: A discoid holdfast supports one or more bright red blades that have a distinct midrib and, in mature thalli, numerous lateral veins. The blades erode during the winter and leave the midrib and veins as a "skeleton" that bear the reproductive structures. New blades may grow on the remnant midrib the next spring.
Pacific Coast Distribution: Pavlof Bay (Alaska Peninsula), Alaska, to Santa Barbara, California.
Habitat: On rocks, in the low intertidal.
Remarks: Some phycologists believe that *Erythrophyllum* found growing epiphytically on kelps is a separate species, *Erythrophyllum splendens*.

Family Petrocelidaceae

Mastocarpus jardinii (J. Agardh) J. West
Description: A reddish-brown, 5–10 cm tall, thallus with narrow (5–10 mm wide) blades that are more or less dichotomously divided. The sides of the blades have a thickened ridge and the blades can have groups of bumps (papillae).
Pacific Coast Distribution: Langara Island (Queen Charlotte Islands), British Columbia, to Santa Barbara County, California.
Previous names: Gigartina agardhii Setchell *et* Gardner; *Gigartina jardinii* J. Agardh
Habitat: On rocks, in the mid to high intertidal.

Mastocarpus papillatus (C. Agardh) Kützing
Turkish Washcloth

Description: The thallus consists of purplish-black blades that are usually covered with knobby pumps (called papillae) that are associated with the reproductive structures. Blades can be narrow (1–2 cm) or wide (5–7 cm) and are dichotomously divided.
Pacific Coast Distribution: Bering Sea and Aleutian Islands, Alaska, to Punta Baja, Baja California.
Previous names: Gigartina papillata (C. Agardh) J. Agardh
Habitat: Abundant, on rocks, in the mid to high intertidal.
Remarks: The morphology of this species is extremely variable, thin bladed forms are easily confused with *Mastocarpus jardinii.*

Mastocarpus exhibits heteromorphic alternation of generations. The upright thalli are the gametophytes and they alternate with soft, black crust. Before the life cycle was known, phycologists thought the tetrasporophyte crust was a distinct species in the genus *Petrocelis*.

Family Phyllophoraceae

Ahnfeltiopsis linearis (C. Agardh) Silva *et* DeCew
Description: A regularly dichotomously branched, brownish-maroon thallus, 10–18 cm tall, that is attached to the substrate via a small disk-shaped holdfast. The branches are cylindrical near the bottom of the thallus and become flattened after the first dichotomy.
Pacific Coast Distribution: Barkley Sound (Vancouver Island), British Columbia, to Point Conception, California.
Previous names: Gymnogongrus linearis (C. Agardh) J. Agardh.
Habitat: Common, on rocks, partially buried with sand, in the mid to low intertidal.
Remarks: Usually found in dense patches where a sandy beach joins a rocky habitat.

Family Solieriaceae

Sarcodiotheca gaudichaudii (Montagne) Gabrielson

Description: The thallus, which attaches to the substrate with a discoid holdfast, is composed of cylindrical, fleshy branches that usually have pointed tips. The branches have an irregular, radial branching arrangement. Thalli are red to reddish-brown in color and 10 to 45 cm tall.
Pacific Coast Distribution: Ketchikan, Alaska, to Baja California (also found in the Galápagos Islands; Peru; and Chile).
Previous names: Neoagardhiella gaudichaudii (Montagne) Abbott; *Agardhiella gaudichaudii* (Montagne) Silva *et* Papenfuss; *Agardhiella coulteri* (Harvey) Setchell; *Agardhiella mexicana* Dawson; misapplied name: *Neoagardhiella baileyi* (Kütz) Wynne *et* Taylor
Habitat: Common to abundant, on rocks near sandy habitats, in the low intertidal and subtidal.
Remarks: Another species of red algae, *Gardneriella tubifera*, is an obligate parasite on *Sarcodiotheca*. The parasite's spores germinate on the host and a parasite nucleus, mitochondria and ribosomes are transferred into a host cell. The altered host cell then divides to create the parasitic algal body.

Order Halymeniales
Family Halymeniaceae

Halymenia spp.

Description: A bright to dark-red thallus of one or more oval or tapering blades that attach via a discoid holdfast. The blades, 45–70 cm tall, are soft and slippery.
Pacific Coast Distribution: Northern British Columbia to Baja California.
Habitat: On rocks, low intertidal and subtidal.
Remarks: The members of this genus are easy to confuse with other foliose red algae, particularly those in the genus *Schizymenia*. Accurate identification requires microscopic examination of cross sections of the blades.

Prionitis lanceolata (Harvey) Harvey

Description: The dark-brown to purplish-brown thallus (up to 25 cm tall) consists of several to numerous cylindrical to flattened stipes that exhibit irregular dichotomous branching. Extending from the stipes are flattened, tapering branches that bear numerous, lanceolate bladelets in a pinnate pattern.

Pacific Coast Distribution: Sitka, Alaska, to Isla Cedros, Baja California.

Previous names: *Zanardinula lanceolata* (Harvey) DeToni

Habitat: On rocks, throughout intertidal and upper subtidal.

Remarks: Fresh *Prionitis* has a bleach-like smell, especially if squeezed.

Prionitis lyallii (Harvey) Harvey

Description: A reddish- to yellowish-brown thallus of flattened, lanceolate blades (up to 5 mm wide) that have an irregular branching pattern. Small pinnately arranged bladelets may be present on the margins of the lance-shaped blades.

Pacific Coast Distribution: Southern British Columbia to Isla Magdalena, Baja California.

Previous names: *Zanardinula lyallii* (Harvey) DeToni

Habitat: On rocks, throughout intertidal region and upper subtidal, often in high pools.

Remarks: *Prionitis* exhibits a lot of morphological variability, even at the species level, as a result, there is debate about the taxonomy within this genus.

Family Champiaceae

Gastroclonium subarticulatum in the low intertidal.

Close-up of *G. subarticulatum* branches.

Gastroclonium subarticulatum (Turner) Kutzing

Description: An irregularly dichotomous-branched thallus, reddish brown to yellowish green in color, 10 to 20 cm tall, with cylindrical, somewhat fleshy, branches. In the upper portion of the thallus, the branches bear cylindrical branchlets that are slightly constricted at regular intervals and have blunt, rounded, usually hollow, tips.

Pacific Coast Distribution: Sitka, Alaska, to Punta Abreojos, Baja California.

Previous names: Gastroclonium coulteri (Harvey) Kylin

Habitat: Common, on rocks, other algae and worm tubes, forming a turf in the mid to low intertidal and subtidal.

Remarks: During the winter, the upper portion of the thallus erodes and leaves an irregularly shaped holdfast.

Family Fauchaceae

Fauchea laciniata J. Agardh

Description: The 3-12 cm tall thallus forms a loose clusters of fan-shaped blades that may be highly branched at apices. The blades are deep-red in color and are often a brightly iridescent blue-violet.

Pacific Coast Distribution: Prince William Sound, Alaska, to Baja California.

Previous names: Fauchea laciniata f. *pygmaea* Setchell *et* Gardner; *F. pygmaea* (Setchell *et* Gardner) Kylin; *F. media* Kylin

Habitat: On rocks, very low intertidal and subtidal.

Remarks: The iridescence seen on blades of *Fauchea* is not related to any light producing phenomena (like bioluminescence), but results from reflected and refracted light off a multi-layered cuticle.

Botryocladia pseudodichotoma (Farlow) Kylin Sea Grapes

Description: A striking, rose-red alga with irregular dichotomous branches, each of which terminate in a mucilage-filled, spherical sac.

Pacific Coast Distribution: Queen Charlotte Islands, British Columbia, to Baja California (also known from Galápagos Islands).

Habitat: On rocks and other algae, in the subtidal.

Remarks: The Greek roots of *Botryocladia* are *botrys* and *klados*, which mean 'cluster of grapes' and 'branch' respectively.

Fryeella gardneri (Setchell) Kylin

Description: The thallus is a cluster of dichotomously branched blades, up to 16 cm tall, that are deep-red to orange-red in color and often have a bluish iridescent cast. The thallus has a very thin, hollow central cavity with arched partitions which can be seen on the surface of the blade as concentric arches.

Pacific Coast Distribution: Queen Charlotte Islands, British Columbia, to Baja California.

Previous names: *Fauchea gardneri* Setchell; *Rhodymenia gardneri* (Setchell) Kylin; *Fryeella gardneri* var. *prostrata* Dawson *et* Neushul.

Habitat: Frequent, on rocks, worm tubes and other algae, in the subtidal.

Remarks: This species is named for the phycologist N. L. Gardner who, along with W. A. Setchell, worked out of the University of California Berkeley in the early 1900's and made enormous contributions to our understanding of marine algae.

Rhodymenia pacifica Kylin
Description: The flabellate thallus, 4–15 cm tall, consists of one to many dichotomously divided blades, 4-14 mm wide. The apices of the deep rose-red blades are often lobate and rounded.
Pacific Coast Distribution: Queen Charlotte Islands, British Columbia, to Baja California.
Previous names: Rhodymenia lobulifera Dawson
Habitat: Common, abundant in the southern portion of the range, on rocks, low intertidal and subtidal.

Sparlingia pertusa (Postels *et* Ruprecht) Saunders, Strachan *et* Kraft
Description: The rose red to yellowish red thallus consists of a single blade about 50 cm long and 20 cm wide that narrows down to a small disk-shaped holdfast. The blade is often divided into several broad segments and, except in very small thalli, is perforated with numerous small holes.
Pacific Coast Distribution: Aleutian Islands, Alaska, to Coos Bay, Oregon.
Previous names: Rhodymenia pertusa (Postels *et* Ruprecht) J. Agardh; *Rhodymenia stipitata* Kylin
Habitat: On rocks, often in exposed habitats, low intertidal and subtidal.

Order Gracilariales
Family Gracilariaceae

Gracilaria and *Gracilariopsis*

Description: A reddish-brown to yellowish-brown thallus, up to 1.5 m tall, composed of few to many cylindrical axes that are anchored by a discoid holdfast. Branches may be sparse or dense, have irregular branching pattern, and a cartilaginous texture.
Pacific Coast Distribution: Southeastern Alaska to Gulf of California, Mexico.
Habitat: Common, on rocks and shells, in moderately exposed habitats to sheltered bays and estuaries, in the low intertidal and subtidal.
Remarks: These very similar genera are distinguished by reproductive structures. Both are edible and can be blanched, pickled, or eaten fresh in salads. *Gracilaria* is a very popular sea vegetable in Japan, where it is called "ogo," and Hawaii, where it is known as "limu loa." It is also cultivated for agar production in Asia, South America, and Africa.

Order Plocamiales
Family Plocamiaceae

Plocamium oregonum Doty

Description: The thallus, rarely more than 10 cm tall, is deep red, flattened and bears alternating groups of curved ramuli (or branchlets) that arise from an undulating axis. Distinguished from other species of *Plocamium* by only having one or two branched ramuli between unbranched ramuli.
Pacific Coast Distribution: Strait of Georgia, British Columbia, to Santa Cruz County, California.
Habitat: On rocks, in the low intertidal and subtidal.

Plocamium cartilagineum covering the top of an intertidal boulder.

A close-up of *P. cartilagineum*.

Plocamium cartilagineum (Linnaeus) Dixon

Description: A flattened alga with alternating groups of two to six curved ramuli (or branchlets) that arise from an undulating axis. There are usually three or four branched ramuli between unbranched ramuli. The thallus is pinkish-red to rose-red in color and can reach 25 cm tall.

Pacific Coast Distribution: Southern southeastern Alaska to Baja California (also Chile and the Galápagos Islands).

Previous names: Plocamium coccineum var. *pacificum* (Kylin) Dawson

Habitat: Common, on rocks and other algae, in the mid intertidal to subtidal.

Remarks: *Plocamium cartilagineum* is widely distributed in the Pacific, Atlantic and Indian Oceans and some phycologists consider the population found along our coast to be a separate species, *Plocamium pacificum* Kylin. Another species found here, *Plocamium violaceum*, has a similar branching pattern but is much smaller (usually only 2–3 cm tall).

A tiny (5 mm in diameter), white thalli can sometimes be found on the branches of *Plocamium*. This is separate species, *Plocamiocolax pulvinata*, which is an obligate parasite and acts by injecting its nuclei and mitochondria into the host cells. It has been proposed that *P. pulvinata* evolved from *Plocamium violaceum*.

PHYLUM ANTHOPHYTA
Class Monocotyledones

Order Najadales
Family Potomogetonaceae

Phyllospadix spp. Surfgrass

Description: The leaves, 1–3 m long, are grass-green to bright green in color and generally less than 4 mm wide.
Pacific Coast Distribution: Alaska to Baja California.
Habitat: On rocks, in wave-exposed areas, lower intertidal and subtidal.
Remarks: There are at least three species of *Phyllospadix* found along our coast. The species are differentiated by characteristics such as the number of roots that arise from each rhizome, blade width, and geographic distribution.

Family Zosteraceae

Zostera spp.
Eelgrass

Description: The dark green leaves are ribbon-like, 1.5–12 mm wide (usually wider than 5 mm) and can be 3 m in length
Pacific Coast Distribution: Alaska to Baja California.
Habitat: Common, on mud or sand in protected waters of bays and estuaries, low intertidal and subtidal.
Remarks: Eelgrass beds create important estuarine habitat for many species including: juvenile fish, crabs and clams.

A non-native species, *Zostera noltii,* has been introduced from Europe and can be locally abundant in northern Washington and southern British Columbia.

Seaweeds are versatile and can be prepared in elaborate dishes or in simple soups and teas. We have provided a few recipes to start you on your adventure of cooking with sea vegetables, but be creative! We have emphasized recipes that use seaweeds that are available both locally and commercially. Comments about additional edible species are included throughout the book. There are several excellent cookbooks that specialize in seaweeds. A list of some of these is provided in the references section.

Laminaria Soup Stock (dashi)

Dried *Laminaria* fronds are sold as kombu. If you harvest fresh *Laminaria*, preserve it by cutting the fronds into pieces and sun drying them.

1 ounce dried *Laminaria* fronds (kombu)
1 ounce dried bonito flakes (katsuobushi), optional
4 cups water

Place kelp and water in a sauce pan. Slowly heat to almost boiling, but do not boil (the stock will become slightly bitter if it boils). Simmer for 5 minutes. Remove from heat, add the bonito flakes (this soup traditionally contains bonito flakes, but they can be omitted if you do not have any available or would like to make a vegetarian stock). Strain; discard the bonito and the kelp.

This delicate stock forms the basis for many Japanese soups and sauces. Here are two ideas:
§ Add 1 lb. of white meat fish fillets, cut into 1 inch pieces. Bring broth briefly to a boil.
§ Add 1 tsp of miso for each cup of broth. Garnish with chopped green onions, a splash of hot oil, or lemon juice.

Udon Soup with Chicken and *Alaria*

Strictly speaking, wakame is a species of seaweed found in Japan called *Undaria*, but in North America, *Alaria* is also sold as wakame. If you harvest *Alaria*, cut the blade well above the reproductive sporophylls and sun dry it.

6 cups *Laminaria* soup stock
chopped stems from 1 bunch of cilantro
chopped stems from 1 bunch of parsley
2 inches of fresh ginger loosely chopped
1 package of dried shitake mushrooms
splash of soy
1 tsp fresh garlic, minced
½ tsp fresh ginger, minced
2 boneless, skinless chicken breasts cut into pieces
½ cup thinly sliced white onion
½ cup thinly sliced carrots about 2 inch long
1 cup dried *Alaria* cut in strips (use scissors)
1 package of fresh, Japanese style noodles
2 cups loosely chopped fresh spinach
1-4 green onions chopped

Combine first 6 ingredients (*Laminaria* stock through soy). Simmer 30 minutes. Strain, and save the broth and the shitakes. Saute next 6 ingredients (garlic through *Alaria*) in sesame oil for a few minutes, then add the mixture to the broth. Remove the woody stems from the shitake mushrooms, slice the caps and add them to the broth. Add the spinach and green onions. Meanwhile, cook noodles according to direction on the package. Place a portion of noodles into each bowl and cover them with soup. Garnish with fresh cilantro and a dash of hot oil.

Corn & *Palmaria* Chowder

Palmaria, or dulse, was traditionally harvested in the North Atlantic. Dried pieces can be snipped and added as a garnish or the blades can be soaked in water and eaten raw or cooked. The plant is quite delicate, however, and can become mushy, so don't overcook it. *Palmaria* works especially well in recipes with potatoes.

2 slices bacon
1 medium onion, chopped
2 14-oz cans vegetable broth
1 bay leaf
1 fresh thyme sprig
2 medium russet potatoes
2 cups frozen corn
1 cup cream
½ cup dried *Palmaria* (dulse)
salt and freshly ground pepper
½ Tbs finely chopped parsley

In a heavy bottomed 2-quart pot, cook bacon for 10 minutes until the fat is rendered. Remove the bacon and chop into small pieces, set aside. Turn down heat and add onions, stirring regularly to keep from browning. When the onions soften and turn translucent (about 10 minutes) add vegetable broth, bay leaf, and thyme. Peel and cut potatoes into ¼ inch cubes and add to soup. Simmer soup for about 10 minutes to cook potatoes. Rehydrate the *Palmaria* in 2 cups of water, then drain and chop into small pieces. Add seaweed and remaining ingredients. Simmer for about 1 minute to warm corn. Remove 1 cup of solids from soup, blend in blender until creamy, add back into soup. Mix well and serve.

Variation:
§ Other algae can easily be added to this recipe. *Alaria*, *Halosaccion* or *Scytisiphon* make excellent variations and additions.

Seaweed & Sushi-Rice Salad

This salad is an excellent side dish for grilled salmon or seared tuna (sushimi)

2 cups water
2 cups sushi rice
½ cup rice vinegar
2 Tbs sesame oil
1 Tbs soy sauce
2 tsp minced fresh ginger
4 tsp minced fresh garlic
¼ - ½ tsp wasabi, depending on taste
1 cup wakame (dried *Alaria* or *Undaria*), optional
chopped cilantro
green onions chopped
sliced veggies (carrots, peppers, snow peas, etc.)
1 Tbs sesame seeds
1 cup toasted nori sliced thin strips

Bring water to boil in medium saucepan; add rice. Cover with tightly fitting lid, reduce heat, and simmer for 20 minutes or until all the water is absorbed. Remove from heat, place in wood or glass salad bowl. Fan and gently fold the rice to cool it to room temperature. Place ingredients for dressing (vinegar through wasabi) in a small bowl and mix well. Reconstitute wakame in water for 15 minutes, squeeze out excess water and chop. Add dressing, wakame, cilantro, green onions and your choice of other vegetables to the rice and toss. Garnish with the sesame seeds and sliced nori.

Alaria & Cucumber Salad

½ cup dried *Alaria* (wakame) cut into strips
2 Tbs soy
3 Tbs cider vinegar
1 Tbs honey
½ tsp sesame oil
1 peeled, seeded and thinly sliced cucumber
½ tsp fresh garlic, minced
2 green onions, chopped
2 Tbs fresh cilantro, chopped

Place the *Alaria* in a bowl, cover with warm water, and let soak for about 20 minutes. Combine soy, vinegar, honey and sesame oil into small jar and shake until well mixed (you may need to heat the dressing to melt the honey, then cool it to room temperature). Drain the seaweed, rinse with cold water, and squeeze out excess water. Combine the seaweed and the rest of the ingredients in a salad bowl, pour the dressing over the top and toss. Refrigerate 30 minutes before serving. Garnish with roasted sesame seeds.

Variations:
§ Add 1 ounce bean thread noodles, cooked according to directions on package.
§ Add baby carrots sliced julienne and/or thinly sliced red pepper

Coleslaw with *Palmaria*

Adapted from a recipe in Lesley Ellis' seaweed cookbook.

¼ cup dried *Palmaria* (dulse)
¼ cup mayonnaise
2 Tbs plain yogurt
1 Tbs red wine vinegar
1 tsp lemon juice
pepper
tarragon
½ cup raisins
¼ cup pine nuts
½ an apple diced
3 cups cabbage, chopped
2 green onions, chopped

Soak the seaweed in water for about 5 minutes. Drain, squeeze out remaining water, chop, and set aside. Combine the next 4 ingredients (mayo through lemon juice) in a salad bowl and mix. Season with black pepper and tarragon. Add the chopped *Palmaria* and the remaining ingredients and mix.

Black Beans with Seaweed

2 cups dried black beans
2 cups water
1 cup apple cider
1 cup red wine
1 can crushed tomatoes
2 Tbs olive oil
1 piece dried *Laminaria* (kombu) snipped into pieces
1 bay leaf
1 carrot, diced
1 onion, chopped
3-4 cloves garlic, minced
1 tsp each of cumin, basil, rosemary, and oregano
¼ cup dried hijiki, reconstituted

Place beans in a large pot. Add the next 7 ingredients (water through bay leaf). Bring to a boil, let simmer 1½ hours. Stir occasionally and, if necessary, add more liquid. In a medium skillet cook the carrots, onion, garlic, spices, and hijiki until the vegetables are tender. Add the mixture to the beans and let simmer another 10 minutes or until the beans are tender. Garnish with sour cream and serve with salsa, rice, and corn tortillas.

Tofu & Hijiki Stir Fry

Hijiki (*Hijikia fusiforme*) is a richly flavored seaweed that is sold dried. After being soaked in water to reconstitute, it is used in soups, salads and stir fries. The tender tips *Cystoseira* (a close relative of *Hijikia*) or fresh *Pelvetiopsis* are local substitutes for this dish.

½ cup dried hijiki
¼ cup vegetable oil
1 block (14 oz.) firm tofu
1 tsp fresh garlic, minced
½ tsp fresh ginger, minced
¼ of an onion sliced into half moons
1 cup baby carrots sliced julienne
2 cups broccoli florets
2 tsp sesame oil

Cooking sauce:
1 tsp cornstarch
1 Tbs soy
1 Tbs peanut sauce
½ cup chicken broth

Wash the hijiki, place it in a small bowel, cover it with water and let soak for about 10 minutes or until it is soft. Bring 3 cups of water to a boil and add the reconstituted hijiki, boil for 5 minutes. Drain hijiki into a colander, rinse with cold water, and set aside. Slice the tofu into ½ inch cubes. Preheat the oil in a wok, as soon as it is smoking slightly, add the tofu and fry for 5-10 minutes until golden brown. Remove tofu and drain on a paper towel. Combine the cooking sauce ingredients and set aside. Remove all but 2-3 Tbs of oil from the wok and reheat until smoking. Add the garlic and ginger and let cook for a few seconds. Add vegetables and hijiki and stir fry briefly. Add a spoonful of chicken broth, cover wok, and let steam briefly. Add the tofu. Stir cooking sauce and add it to the wok. Cook for about 30 seconds. Remove stir fry from heat. Garnish with the sesame oil. Serve hot with rice.

Sushi

3 cups water
2 cups sushi rice
1 2-inch long piece dried *Laminaria*, optional
2 Tbs rice vinegar
2 Tbs sugar
1 tsp salt
5 sheets toasted nori (processed and dried *Porphyra*)

Filling ideas (be creative!):
sliced carrots, cucumber, green onions, avocado, etc
shrimp and/or crab meat
smoked salmon
fresh, raw tuna
scrambled egg

Place water in 2 quart saucepan with a tightly fitting lid. Add the rice and place the kombu on top. Bring the rice to a boil. Reduce heat, cover, and simmer until water is absorbed (about 15 minutes). Remove from heat, discard kombu, and let rice sit for 5 minutes. While the rice is cooking, put the vinegar, sugar and salt into a small saucepan and mix thoroughly. Heat slowly, just long enough for the sugar and salt to dissolve. Remove from heat and let cool. While the rice is still very hot, transfer it to a wooden or glass bowl and add the dressing. Toss and fold the rice (do not stir or mash it) with a wooden paddle or spoon. While you are tossing the rice, fan it to cool it quickly.

Place the nori on a bamboo mat, if you have one. Wet your hands and cover the nori with about 1 cup of rice in an even layer. Leave a 1-inch wide border on the far edge. Add the filling in a strip about 2-inches from the near edge. Roll the nori toward the far edge, moisten the edge to seal. Cut into 1-inch slices with a sharp knife. Repeat with remaining ingredients. Serve right away with wasabi, soy sauce, and pickled ginger.

Glossary

ALTERNATION OF GENERATIONS: The succession of haploid (n) and diploid (2n) thalli in the life cycle of some sexually reproducing organisms, notably plants and many algae.

ANNUAL: Referring to a plant that completes its life cycle in one growing season (contrast with perennial).

APICAL: At or near an apex or tip.

AXIS (pl. axes): The main longitudinal stem which often bears branches or other structures.

BIFLAGELLATE: Bearing two flagella.

BLADELET: A small blade.

BRANCHLET: A small branch.

BULLAE: Blister-like bulges or protrusions on the surface.

BULLATE: Having bullae (blister-like bumps).

CARPOGONIUM (pl. carpogonia): The female sex cell: found in most red algae.

CARPOSPOROPHYTE: A multicellular diploid phase that lives on or in the female gametophyte and produces carpospores: occurs in the life cycle of most red algae.

CARRAGEENAN: A polysaccharide found in the cell walls of red algae; extracted and used in pharmaceutical and food industries.

CHLOROPHYLL: A green pigment associated with the chloroplast, involved in photosynthesis.

CHLOROPLAST: A double membrane-bound organelle that contains photosynthetic pigments.

CONCEPTACLE: A cavity within the thallus that contains reproductive structures.

CRUSTOSE: Forming a crust or crust-like layer on the substrate.

CYSTOCARP: A red algal structure consisting of the carposporophyte and the surrounding gametophyte tissue (pericarp).

DECIDUOUS: Falling off or shed, therefore not always present.

DENTATE: Tooth-like.

DICHOTOMOUS: Divided into uniform pairs of equal length.

DIPLOID: Containing two sets of chromosomes (2n) in each nucleus.

DISCOID: Rounded or disk-like in shape.

EPIPHYTE: Growing upon a plant or alga (generally non-parasitic).

FILAMENTOUS: Thread-like or hair-like; often consisting of a single row of cells.

FIMBRIA (pl. fimbriae): A fringe along an edge.

FLAGELLUM (pl. flagella): A whip-like projection whose beating movement causes movement of the cell.

GAMETANGIUM (pl. gametangia): The structure that produces gametes.

GAMETE: A sexual reproduction cell. Eggs and sperm are examples of gametes.

GAMETOPHYTE: The haploid thallus, that produces gametes.

GLOBULAR: Round or spherical in appearance.

HAPLOID: Containing a single set of chromosomes (n) in each nucleus. Gametes, for example, are haploid.

HAPTERON (pl. haptera): Root-like structure that anchors an alga to the substrate.

HERBIVORY: Feeding upon plants or algae.

HETEROMORPHIC: Having different shapes; when the gametophyte and sporophyte have different forms or appearance.

HOLDFAST: A structure that anchors an alga to the substrate.

INTERCALARY: Occurring in the middle of a structure rather than the apex.

INTERGENICULUM: The highly calcified portion of an articulated coralline algae occurring between the joints.

ISOGAMY: Reproduction where gametes are identical in size and shape.

ISOMORPHIC: Having identical shape; when the gametophyte and sporophyte are similar in form or appearance.

LANCEOLATE: Lance-shaped; narrow and tapering at both ends.

LEAFLET: A small leaf.

LOBATE: Divided into lobe-like shapes.

MEIOSIS: Division of a cell nucleus during sexual reproduction in which one diploid cells produces four haploid daughter cells.

MERISTEM: Dividing tissue that produces growth.

MIDRIB: A distinct line, often thickened, running longitudinally in a blade.

MITOSIS: Division of a cell nucleus which produces identical nuclei in the daughter cells.

OOGAMY: Reproduction in which gametes consist of a large non-motile egg and a smaller sperm cell.

PAPILLA (pl. papillae): A small bump or projection on the surface.

PARENCHYMATOUS: Tissue containing large, thin-walled cells of equal diameter; not filamentous.

PERENNIAL: A plant that lives from year to year.

PERICENTRAL: Around a central axis.

PHOTOSYNTHESIS: A process in which carbon dioxide, water, and the energy of sunlight produce a sugar.

PHYCOLOGY: The study of algae.

PINNATE: Feather-like branching pattern with a main axis and lateral branches on two sides.

PNEUMATOCYST: A large gas-filled structure that serves as a float.

POLYPHYLETIC: A group of organisms that arose from more than one ancestor.

POLYSACCHARIDE: A chain made of many smaller sugar molecules.

PROSTRATE: Lying flat on the substrate.

PSEUDOPARENCHYMATOUS: A tissue composed of densely, interwoven, continuous filaments.

QUADRIFLAGELLATE: Bearing four flagella.

RAMULUS (pl. ramuli): A small secondary branch.

RECEPTACLE: An inflated tip of a branch containing conceptacles.

RHIZOID: A root-like structure or filament that functions in attachment.

RHIZOME: Algal rhizomes are horizontal portions of the thallus that produce additional upright axes. In vascular plants, a rhizome is a specialized underground stem that often grows horizontally.

SACCATE: Sac-like in appearance.

SORUS (pl. sori): A patch or cluster of reproductive structures.

SPECIFIC EPITHET: The second word in a species name; sometimes called the trivial name.

SPERMATANGIUM (pl. spermatangia): A structure that produces a spermatium.

SPERMATIUM (pl. spermatia): A male reproductive gamete with no flagella; in red algae.

SPORANGIUM (pl. sporangia): A cell that produces spores internally.

SPORE: A cell that is able to germinate into a new individual without syngamy.

SPOROPHYLL: A blade-like structure that bears sporangia.

SPOROPHYTE: The diploid thallus, that produces spores, in the life cycle of algae and plants with alternation of generations (contrast with gametophyte).

STIPE: The stem-like portion of the thallus between the holdfast and the blade(s).

TETRASPORANGIUM (pl. tetrasporangia): A red algal structure that produces four spores via meiosis.

TETRASPOROPHYTE: The diploid phase in red algal life cycles that produce tetraspores.

THALLUS (pl. thalli): The body of an alga.

TRICHOTHALLIC GROWTH: Intercalary growth producing a psedoparenchamatous thallus on one side and a filament on the other.

UNIFLAGELLATE: Bearing one flagellum.

UNISERIATE: A single row.

VEGETATIVE: The non-reproductive portions of a thallus.

ZOOSPORE: A motile spore with flagella.

ZYGOTE: The cell that is created from the fusion two of gametes.

Selected References

Abbott, I. A. 1978. Morphological and taxonomic observations on *Neoagardhiella* (Gigartinales, Rhodophyta), with emphasis on Pacific populations. J. Phycol. 14: 48-53.

Abbott, I. A. and G. J. Hollenberg. 1976. Marine algae of California. Stanford, University Press, Stanford, CA, 827 pp.

Aguilar-Rosas, L. E., R. Aguilar-Rosas, A. C. Mendoza-González, and L. E. Mateo-Cid. 2000. Marine algae from the northeast coast of Baja California, México. Bot. Mar. 43:127-139.

Blair, S. M. 1983. Taxonomic treatment of the *Chaetomorpha* and *Rhizoclonium* species (Cladophorales; Chlorophyta) in New England. Rhodora 85:175-211.

Blair, S. M., A. C. Mathieson, and D. P. Cheney. 1982. Morphological and electrophoretic investigations of selected species of *Chaetomorpha* (Chlorophyta; Cladophorales). Phycologia 21(2):164-172.

Bold, H. C. and M. J. Wynne. 1985. An introduction to the algae. Prentice-Hall, Englewood Cliffs, NJ, 720 pp.

Brostoff, W. N. 1988. Taxonomic studies of *Macrocystis pyrifera* (L.) C. Agardh (Phaeophyta) in southern California: holdfasts and basal stipes. Aquat. Bot. 31:289-305.

Chapman, A. R. O. 1972. Morphological variation and its taxonomic implications in the ligulate members of the genus *Desmarestia* occurring on the west coast of North America. Syesis 5:1-20.

Chapman, A. R. O. 1972. Species delimination in the filiform, oppositely branched members of the genus *Desmarestia* Lamour. (Phaeophyceae, Desmarestiales) in the northern hemisphere. Phycologia 11(3/4):225-231.

DeCew, T. C. and J. A. West. 1981. Investigations on the life histories of three *Farlowia* species (Rhodophyta: Cryptonemiales, Dumontiaceae) from Pacific North America. Phycologia 20(4):342-351.

Druehl, L. D. 1990. Molecular evolution in the Laminariales: a review. Pp. 205-217, In: D. J. Garbary and G. R. South (eds.), Evolutionary biogeography of the marine algae of the North Atlantic. Springer-Verlag: Berlin.

Druehl, L. D. 1978. The distribution of *Macrocystis integrifolia* in British Columbia as related to environmental parameters. Can. J. Bot. 56:69-79.

Druehl, L. D. 1968. Taxonomy and distribution of northeast Pacific species of *Laminaria*. Can. J. Bot. 46:539-547.

Druehl, L. D. and L. Kemp. 1982. Morphological and growth responses of geographically isolated *Macrocystis integrifolia* populations when grown in a common environment. Can. J. Bot. 60:1409-1413.

Fredericq, S., M. H. Hommersand, and D. W. Freshwater. 1996. The molecular systematics of some agar- and carrageenan-containing marine red algae based on *rbc*L sequence analysis. Hydrobiologia 326/327:125-135.

Gabrielson, P. W. 1982. Morphological studies of the members of the tribe Agardhielleae (Solieriaceae, Rhodophyta) II. *Sarcodiotheca gaudichaudii* (Montagne) comb. nov. Phycologia 21(1):86-96.

Gabrielson, P. W. and R. F. Scagel. 1989. The marine algae of British Columbia, northern Washington, and southeast Alaska: division Rhodophyta (red algae), class Rhodophyceae, order Gigartinales, families Caulacanthaceae and Plocamiaceae. Can. J. Bot. 67:1221-1234.

Gabrielson, P. W., R. F. Scagel, and T. B. Widdowson. 2000. Keys to the Benthic marine algae and seagrasses of British Columbia, southeast Alaska, Washington and Oregon. Phycological Contribution No. 5, Department of Botany, University of British Columbia, Vancouver, BC, 189 pp.

Gerard, V. A. 1990. Ecotypic differentiation in the kelp *Laminaria saccharina*: phase-specific adaptation in a complex life cycle. Mar. Biol. 104:519-528.

Gerwick, W. H., M. A. Roberts, P. J. Proteau and J.-L. Chen. 1994. Screening cultured marine microalgae for anticancer-type activity. J. Appl. Phycol. 6:143-149.

Goff, L. and I. McLaughlin. 1997. The complete sequence of the mitochondrial genome of the red algal parasitc *Plocamiocolax pulvinata* (Plocamiales) and comparison to its host, *Plocamium cartilagineum*, and two other *Plocamium* non-host species. Phycologia 36(4):35.

Graham, M. H. 1997. Factors determining the upper limit of giant kelp, *Macrocystis pyrifera* Agardh, along the Monterey Peninsula, central California, USA. J. Exp. Mar. Biol. Ecol. 218:127-149.

Guiry, M. D. and E. Nic Dhonncha. 2001. Algaebase [database on the internet]. Environmental Change Institute, Galway, Ireland. Available online at http://www.algaebase.org.

Hansen, G. I. 1997. A revised checklist and preliminary assessment of the macrobenthic marine algae and seagrasses of Oregon. Pp. 175-200 *in* T. N. Kaye, A. Liston, R. M. Love, D. L. Luoma, R. J. Meinke, and M. V. Wilson [Eds.], *Conservation and Management of Native Flora and Fungi*. Native Plant Society of Oregon, Corvallis, OR.

Harrison, P. G. and R. E. Bigley. 1982. The recent introduction of the seagrass *Zostera japonica* Aschers. and Graebn. to the Pacific coast of North America. Can. J. Fish. Aquat. Sci. 39:1642-1648.

Hawkes, M. W., C. E. Tanner, and P. A. Lebednik. 1978. The benthic marine algae of northern British Columbia. Syesis 11:81-115.

Hommersand, M. H., S. Fredericq, D. W. Freshwater. 1994. Phylogenetic systematics and biogeography of the Gigartinaceae (Gigartinales, Rhodophyta) based on sequence analysis of *rbc*L. Bot. Mar. 37:193-203.

Hommersand, M. H., S. Fredericq, D. W. Freshwater, and J. Hughey. 1999. Recent developments in the systematics of the Gigartinaceae (Gigartinales, Rhodophyta) based on *rbc*L sequence analysis and morphological evidence. Phycol. Res. 47:139-151.

Hommersand, M. H., M. D. Guiry, S. Fredericq, and G. L. Leister. 1993. New perspectives in the taxonomy of the Gigartinaceae (Gigartinales, Rhodophyta). Hydrobiologia 260/261:105-120.

Hughey, J. R., P. C. Silva, and M. H. Hommersand. 2001. Solving taxonomic and nomenclatural problems in Pacific Gigartinacaea (Rhodophyta) using DNA from type material. J. Phycol. 37:1091-1109

Kogame, K. 1997. Sexual reproduction and life history of *Petalonia fascia* (Scytosiphonales, Phaeophyceae). Phycologia 36(5):389-394.

Kozloff, E. N. 1993. Seashore life of the northern Pacific coast. University of Washington Press, Seattle, WA, 370 pp.

Lewis, R. J. and M. Neushul. 1994. Northern and southern hemisphere hybrids of *Macrocystis* (Phaeophyceae). J. Phycol. 30:346-353.

Lüning, K. and I. tom Dieck. 1990. The distribution and evolution of the Laminariales: North Pacific - Atlantic relationships. Pp. 187-204, In: D. J. Garbary and G. R. South (eds.), Evolutionary biogeography of the marine algae of the North Atlantic. Springer-Verlag, Berlin.

Maggs, C. A., J. L. McLachlan, and G. W. Saunders. 1989. Infrageneric taxonomy of Ahnfeltia (Ahnfeltiales, Rhodophyta). J. Phycol. 25:351-368.

Maggs, C. A, and C. M. Pueschel. 1989. Morphology and development of *Ahnfeltia plicata* (Rhodophyta): proposal of Ahnfeltiales ord. nov. J. Phycol. 25:333-351.

Mahasneh, I., M. Jamal, M. Kashashneh, and M. Zibdeh. 1995. Antibiotic activity of marine algae against multi-antibiotic resistant bacteria. Microbios 83(334):23-26.

Miller, K. A. and H. W. Dorr. 1994. Natural history of mainland and island populations of the deep water Elk Kelp *Pelagophycus* (Laminariales, Phaeophyta): How many species? Pp. 59-70 in W. L. Halvorson and G. J. Maender [Eds.], The Fourth California Islands Symposium: Update on the Status of Resources. Santa Barbara Museum of Natural History, Santa Barbara, CA.

Nam, K. W., C. A. Maggs, and D. J. Garbary. 1994. Resurrection of the genus *Osmundea* with an emendation of the generic delineation of *Laurencia* (Ceramiales, Rhodophyta). Phycologia 33(5):384-395.

Neushul, M. 1990. Antiviral carbohydrates from red algae. Hydrobiologia, 204/205:99-104.

Nisizawa, K., H. Noda, R. Kikuchi, and T. Wanatabe. 1987. The main seaweed foods in Japan. Hydrobiologia 151/152:5-29.

O'Clair, R. M. and S. C. Lindstrom. 2000. North Pacific Seaweeds. Plant Press, Friday Harbor, WA, 161 pp.

Peters, A. F., M. J. H. van Oppen, C. Wiencke, W. T. Stam, and J. L. Olsen. 1997. Phylogeny and historical ecology of the Desmarestiaceae (Phaeophyceae) support a southern hemisphere origin. J. Phycol. 33:294-309.

Phillips, J. A. and M. N. Clayton. 1993. Comparative flagellar morphology of spermatozoids of the Dictyotales (Phaeophyceae). Eur. J. Phycol. 28(2):123-127.

Renfrew, D. E., P. W. Gabrielson, and R. F. Scagel. 1989. The marine algae of British Columbia, northern Washington, and southeast Alaska: division Rhodophyta (red algae), class Rhodophyceae, order Gelidiales. Can. J. Bot. 67:3295-3314.

Saunders, G. W., I. M. Strachan, and G. T. Kraft. 1999. The families of the order Rhodymeniales (Rhodophyta): a molecular-systematic investigation with a description of Faucheaceae fam. nov. Phycologia 38(1):23-40.

Saunders, G. W. and J. C. Bailey. 1997. Phylogenesis of pit-plug-associated features in the Rhodophyta: inferences from molecular systematic data. Can. J. Bot. 75:1436-1447.

Scagel, R. F., D. J. Gabrielson, D. J. Garybary, L. Golden, M. W. Hawkes, S. C. Lindstrom, J. C. Oliveira, and T. B. Widdowson. 1993. A synopsis of the benthic marine algae of British Columbia, southeast Alaska, Washington and Oregon. Phycological Contribution No. 3 Department of Botany: University of British Columbia, Vancouver, BC, 535 pp.

Stein, J. R. and C. A. Borden. 1984. Causative and beneficial algae in human disease conditions. Phycologia 23(4):485-501.

Steneck, R. S. and R. T. Paine. 1986. Ecological and taxonomic studies of shallow-water encrusting Corallinaceae (Rhodophyta) of the boreal northeastern Pacific. Phycologia 25(2):221-240.

Silva, P. 2001. Index Nominum Algarum [database on the internet]. University Herbarium, University of California, Berkeley. Available online at http://ucjeps.herb.berkeley.edu/INA.html.

Taylor, P. R. and M. E. Hay. 1984. Functional morphology of intertidal seaweeds: adaptive significance of aggregate vs. solitary forms. Mar. Ecol. Prog. Ser. 18:295-302.

Trowbridge, C. D. 1992. Mesoherbivory: the ascoglossan sea slug *Placida dendritica* may contribute to the restricted distribution of its algal host. Mar. Ecol. Prog. Ser. 83:207-220.

VanBlaricom, G. R., D. C. Reed, C. Harrold, and J. L. Bodkin. 1986. A sublittoral population of *Pleurophycus gardneri* Setchell and Saunders 1900 (Phaeophyceae: Laminariaceae) in central California. Bull. South. Calif. Acad. Sci. 85(2):120-122.

Vandermeulen, H., R. E. Dewreede, and K .M Cole. 1984. Nomenclature for three species of *Colpomenia* (Scytosiphonales, Phaeophyta). Taxon 33:324-329.

Ver Steeg, J. and M. N. Josselyn. 1983. Taxonomic and morphological studies of *Cryptopleura* (Rhodophyta: Delesseriaceae). Wasman J. Biol. 41(1-2): 97-107.

Widdowson, T. B. 1973. The marine algae of British Columbia and Northern Washington: Revised list and keys. Part I. Paeophyceae (brown algae). Syesis 6:81-96.

Widdowson, T. B. 1971. A taxonomic revision of the genus *Alaria* Greville. Syesis 4:11-49.

Wynne, M. J. 1990. Records and notes of Alaska marine algae. III. Contribution University of Michigan Herbarium 17:335-343.

Wynne, M. J. 1987. Records and notes on Alaska marine algae. II. Contrib. Univ. Michigan Herb. 16:223-232.

Selected Seaweed Cookbooks:

Ellis, Lesley. 1998. *Seaweed: A Cook's Guide.* Fisher Books, Tucson, AZ, 100 pp.

Lewallen, Eleanor and John Lewallen. 1996. *Sea Vegetable Gourmet Cookbook and Wildcrafter's Guide.* Mendocino Sea Vegetable Company, Mendocino, CA, 127 pp.

Madlener, Judith Cooper. 1981. *The Sea Vegetable Gelatin Cookbook.* Woodbridge Press Publishing Company, Santa Barbara, CA, 154 pp.

McConnaughey, Evelyn. 1985. *Sea Vegetables Harvesting Guide and Cookbook.* Naturegraph Publishers, Happy Camp, CA, 239 pp.

Index